Josiah Coates

1,001 Algebra 1

Practice Problems

Workbook

D1196700

Table of Contents

Preface

Greetings Algebra student. Congratulations on your wise decision to pursue supplemental Algebra 1 practice with this workbook. This book is the culmination of several years of researching the most effective style of teaching math problems.

It's not perfect, but then again, neither are you. But maybe you will be, after finishing this book? With thorough and disciplined practice, these practice problems will allow for your complete mastery of Algebra 1.

For your convenience, I took the liberty to include a few example problems at the beginning of each chapter.

Good luck,

Josiah Coates

Chapter 1 Review

Algebra 1 requires familiarity with exponents and order of operations. Below are a few example problems:

$$4^5 = 4 \times 4 \times 4 \times 4 \times 4 = 1{,}024$$

$$-5^3 = -5 \times -5 \times -5 = -125$$

$$5^4 \times 5^5 = 5^{4+5} = 5^9 = 1{,}953{,}125$$

$$\left(\frac{1}{3}\right)^3 = \frac{1^3}{3^3} = \frac{1}{27}$$

$$(4 \times 2)^4 = 8^4 = 4{,}096$$

$$\frac{3^5}{3^3} = 3^{5-3} = 3^2 = 9$$

$$(2^3)^2 = 2^{3 \times 2} = 2^6 = 64$$

$$4^{-3} = \frac{1}{4^3} = \frac{1}{4 \times 4 \times 4} = \frac{1}{64}$$

$$\sqrt{169} = 13$$

$$\sqrt{200} = \sqrt{2 \times 100} = 10\sqrt{2}$$

Practice Problems

1. $3^5 =$

2. $2^3 =$

3. $-7^5 =$

4. $-4^3 =$

5. $7^7 =$

6. $2^3 \times 2^2 =$

7. $6^2 \times 6^3 =$

8. $4^4 \times 4 =$

9. $\frac{5^2}{5^1} =$

10. $\frac{4^7}{4^4} =$

11. $\frac{5^7}{4^4} =$

12. $\frac{5^4}{4^7} =$

13. $\frac{77^7}{77^5} =$

14. $\frac{467^{98}}{467^{97}} =$

15. $\left(\frac{6}{7}\right)^5 =$

16. $\left(\frac{7}{8}\right)^4 =$

17. $\left(\frac{1}{2}\right)^2 =$

18. $\left(\frac{3}{4}\right)^5 =$

19. $(8^2)^4 =$

20. $(6^3)^2 =$

21. $(8)^4 =$

22. $(8^4)^2 =$

23. $(2^2)^{-2} =$

24. $2^{-1} =$

25. $6^{-2} =$

26. $3^{-2} =$

27. $(8^2)^{-2} =$

28. $\sqrt{25} =$

29. $\sqrt{144} =$

30. $\sqrt{169} =$

31. $\sqrt{100} =$

32. $\sqrt{60} =$

33. $\sqrt{80} =$

34. $\sqrt{17} =$

35. $\sqrt{18} =$

36. $\sqrt{27} =$

37. $(3 \times 4 + 3^2) =$

38. $(3 + 4 \times 3^2) =$

39. $(4 \times 3 + 4^2) =$

40. $(4 + 3 \times 4^2) =$

41. $(4 \times 3 + 4^2) + (4 + 3 \times 4^2) =$

42. $(4 \times 3^2 + 4) + (4 \times 3^2 \times 4) =$

43. $(2 \times 2^2 + 2) \times (3 \times 3^3 + 3) =$

44. $6 \times 3 + (2 \div 1) \times 3 + \frac{3}{4} =$

45. $\frac{(2 \times 2^2 + 2)}{(3 \times 3^3 + 3)} =$

46. $\frac{(2 \times 2^2 + 4)}{\sqrt{2 \times 4 \times 2}} =$

47. $\frac{(4 \times 5^2)}{\sqrt{3 \times 25 + 5^2}} \times \frac{\sqrt{4 \times 50 + 5^2}}{(3 \times 3^2 \div 9)} =$

48. $\dfrac{\left(\frac{1}{2}\right)^3}{\sqrt{4\times15+2^4}} \div \dfrac{\left(8\times7^2\right)}{\left(3\times3^2\div9\right)} =$

50. $\dfrac{2^2}{\left(2\times\frac{1}{3}\right)} - \dfrac{2^{-4}}{5^2} =$

49. $\dfrac{\left(\frac{1}{2}\right)^3}{3^{-2}} + \dfrac{4^{-3}}{\left(3\times\frac{7}{8}\right)} =$

Chapter 2 Functions

Assume the price of gas today is around $2.00 per gallon. If we assume that $2.00 is the input, then 1 gallon of gasoline is the output. If we input $4.00 into the gas pump, then output will be 2 gallons of gas, and so forth. This relationship is demonstrated in the table below:

Input	Output
$2.00	1 gallon
$4.00	2 gallons
$6.00	3 gallons
$8.00	4 gallons
$10.00	5 gallons

This relationship of inputs and outputs is what is known as a function. In general, the input is usually denoted by the variable "x" and the output is denoted by the variable "y."

Using these variables, the relationship between money and gallons is as follows:

$$y = \frac{x}{2}$$

Where y=gallons, and x=dollars.

This relationship can also be demonstrated in the graph below:

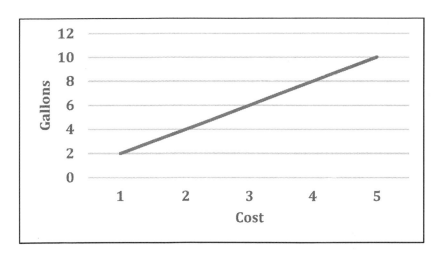

Herein lies the essence of Algebra. Algebra is about the relationship between an input, and a distinct and separate output. The input is independent, and the output is dependent.

There are infinite possibilities of relationships that can be governed by algebraic equations. The number of vaccines that can be created from a certain number of incubators, the health of the economy based on jobs data, the amount of gas a car consumes based on its average speed. These are all examples of relationships that can be expressed in terms of an algebraic equation.

A basic form of equation in Algebra is a linear equation. A linear equation is a type of equation that represents the graph of a straight line. The most common format of a linear equation is shown below:

$$y = mx + b$$

Here, x is the input and y is the output. The letters "m" and "b" represent constants. The "b" denotes the location where the graph intersects the y-axis and "m" denotes the slope of the equation.

Practice Problems

What is the y-intercept of the below equations?

51. $y = x + 1$

56. $3y = 2x + 2$

52. $y = 2x - 1$

57. $x = y - 1$

53. $y = 3x + 2$

58. $-x = y + 1$

54. $y = -4x + 7$

59. $4y = x + 8$

55. $2y = 4x + 2$

60. $y = 5$

What is the slope of the below equations?

61. $y = x + 1$

62. $y = 2x - 1$

67. $x = y - 1$

63. $y = 3x + 2$

68. $-x = y + 1$

64. $y = -4x + 7$

69. $4y = x + 8$

65. $2y = 4x + 2$

70. $y = 5$

66. $3y = 2x + 2$

Write the equation of the below graphs in y-intercept form:

71.

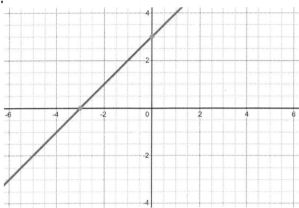

72.

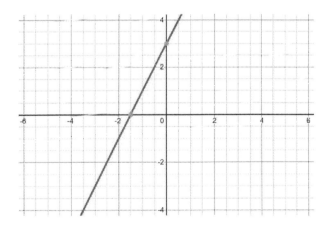

73.

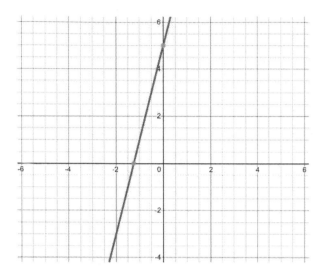

74.

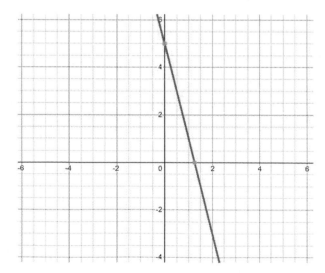

75.

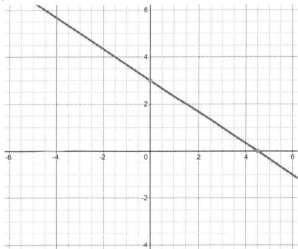

76.

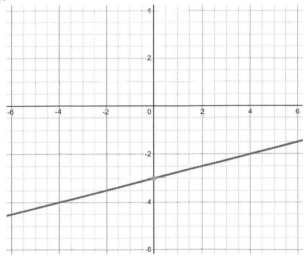

77.

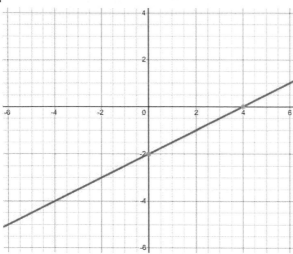

78.

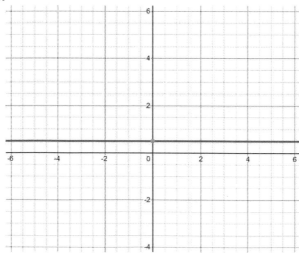

79.

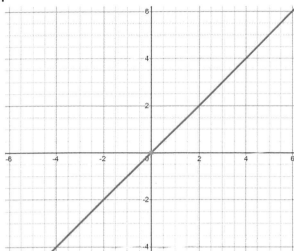

80.

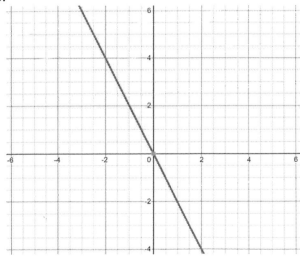

Indicate if the below graphs represent a function:

81.

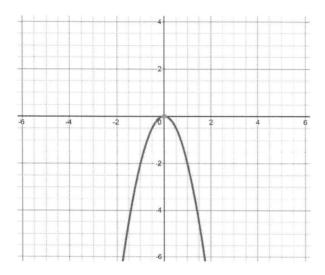

82.

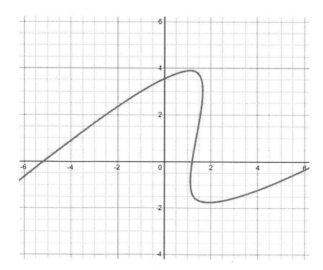

83.

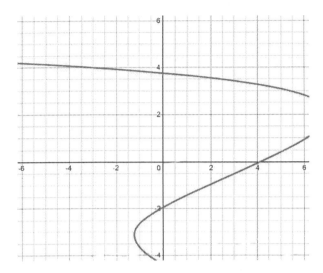

84.

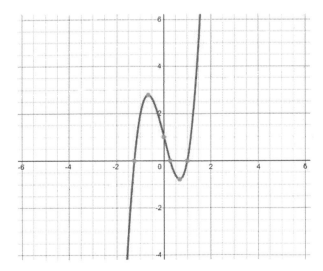

85.

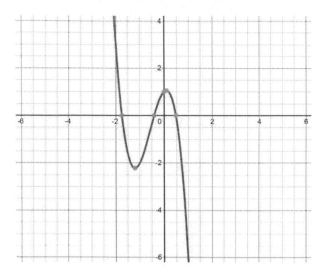

Chapter 3 Solving Equations

When solving equations, we seek to determine the value of the variable for which the equation is true. Below is an example.

Solve for x:

$$4x + 10 = 2x - 4$$

The first step is to combine like terms:

$$4x - 2x = -4 - 10$$

Now we simplify:

$$2x = -14$$

Now we can divide both sides by 2 to solve for x:

$$\frac{2x}{2} = \frac{-14}{2}$$

$$x = -7$$

Practice Problems

Solve for x:

86. $4(4x - 3) = 2(7x - 1)$

88. $2(4x - 4) = 3(8x - 12)$

87. $2(5x - 6) = 8(x - 4)$

89. $4x + 3 = 3x - 4$

90. $7x + 14 = 21x + 28$

98. $9x + 8 = 4x - 3$

91. $x + 2 = 3x - 4$

99. $5x - 5 = 4x + 17$

92. $12 + 2x = 8$

100. $12x + 15 = 15x - 18$

93. $3(x + 1) = 6x - 4$

101. $13(2x - 8) = 7(8x + 7)$

94. $2 - x = x - 2$

102. $15x + 20 = 12x + 25$

95. $3x + 9 = 7 + x$

103. $7x + 14 = 21x + 28$

96. $9x - 9 = 21 + 7x$

104. $13 + 17x = 21$

97. $(3 - 4x)5 = 20 - 15x$

105. $12(4x - 5) = 11(8x - 12)$

106. $14(9x - 7) = 13x + 18$

107. $22(8 - 3x) = 12x + 46$

108. $12(7x - 7) = 20x$

109. $2(4x - 2) = 32x - 2$

110. $8(3x - 9) = 4x + 18$

111. $16 + 6x = 41 + x$

112. $x + 27 = 3x + 32$

113. $18x + 21 = 26x - 35$

114. $19x + 81 = 42x - 34$

115. $42x - 34 = 32 + 15x$

116. $27 - 7x = 3x - 3$

117. $12 + 12x = 36$

118. $x + 1 = 2x + 28$

119. $3(6x - 7) = 2(8x - 14)$

120. $5(2x - 7) = 8(7x - 21)$

121. $12(10x + 10) = 6(5x - 13)$

122. $x + 1 = -x - 2$

123. $3x + 12 = -4x - 9$

124. $7(2x + 6) = 16x - 12$

125. $12x - 20 = 15 - 8x$

126. $(1 - x)12 = 5 - 12x$

127. $51x - 52 = 34x + 16$

128. $7(2x - 4) = 9(6x + 8)$

129. $4x + 24 = 31x + 54$

130. $6(5x - 8) = 12(4x - 8)$

131. $4(22 - 12x) = 12x + 44$

132. $8(7x - 9) = 12(10x - 7)$

133. $4(4x - 7) = 9(9x + 8)$

134. $(7 - x)5 = 9 - 11x$

135. $5(2x + 5) = 3x - 10$

136. $9x + 12 = 8x - 4$

137. $6(3x + 8) = 2(6x + 19)$

138. $3x + 8 = 9x + 27$

146. $32x - 94 = 62 + 35x$

139. $24 + 5x = 8x + 3$

147. $4 - 2x = 6x + 72$

140. $12x - 30 = 46x + 38$

148. $2(6x + 9) = 2(3x + 16)$

141. $2x + 18 = 6x + 24$

149. $2(20x + 30) = 4(4x - 12)$

142. $12(4x - 4) = 6x + 18$

150. $-24x + 120 = 15x + 90$

143. $3(9x - 6) = 3x - 12$

151. $|4 + x| = 12$

144. $14 + 16x = 42 + 2x$

152. $|2 + x| = 6$

145. $28x + 41 = 56x - 65$

153. $|3 + x| = 9$

154. $2|2 + x| = 8$

158. $8|3 + x| = 24$

155. $4|3 + x| = 16$

159. $2|x + 2| = 12$

156. $|x + 1| = 7$

160. $8|x + 2| = 2$

157. $6|x + 12| = 2$

Combine like terms:

161. $2y + 3x + 4y + x$

162. $y + x - 2y + 3x$

163. $7y + 7x - 3y + 3x$

164. $2y + 3x + 4y + 6x$

165. $2xy + 2y + 2xy$

166. $3xy + 7x + 4xy - y + x$

167. $2xy + 3x - 3y + 4 + 5xy - 12$

168. $2y + 3x - 10y + 4x - 5y - 3y + 3xy - 2x$

169. $x + 6x - 6xy - 5y + 5y - 9x + 8z - 9z$

170. $6y + 5y - 8z - xy - 5y - 6z + xy - 4y$

171. $8xy + 7xy + 6xy - 4 - 5y + 4x - 3x$

172. $3z + 8xy - 7x + 9z + 5y - 4xy - 6$

173. $9xz + 3z + 6y - 4 + 5y - 2x + 9z - 9$

174. $6x - 9xy - 2 + 5y - y + 7xy - 3z$

175. $2 + 3y + 4z - 6 - 5y - 6x + 4z - 2y$

176. $6x + 2z + 2y - 7x + 5xy + 8z + 2xy$

177. $z - 4x + 2xy - x - z - 3xy + 2z$

178. $z + 3xy - y + 5y + 2y - 7xy$

179. $xz + 3z + 6y + 7z + -2xz + y - 2x$

180. $xy + y + 7z - 6 + 8xy - 6z + 2y - 4xy$

Chapter 4 Inequalities

Below is an example of a typical inequality:

$$x + 5 \leq -4$$

$$x \leq -4 - 5$$

$$x \leq -9$$

Some inequalities require division to solve:

$$2x \geq 14$$

$$\frac{2x}{2} \geq \frac{14}{2}$$

$$x \geq 7$$

When we are dividing by a negative number to solve for x, we must change the direction of the sign:

$$-2x \geq 14$$

$$\frac{-2x}{-2} \geq \frac{14}{-2}$$

$$x \leq -7$$

Below is a more complicated inequality, requiring the distributive property to solve.

$$3(5x - 5) + 25 \leq 5(4 + 5x)$$

First follow distributive property:

$$15x - 15 + 25 \leq 20 + 25x$$

Then combine like terms:

$$15x - 25x \leq 20 + 15 - 25$$

Simplify:

$$-10x \leq 10$$

Divide to solve for x:

$$\frac{-10x}{-10} \leq \frac{10}{-10}$$

$$x \geq -1$$

Practice Problems

Solve for x:

181. $x + 3 > 2$

182. $x + 7 < 4$ **187.** $x + 6 > 8$

183. $x - 4 < -7$ **188.** $-5 + x < -1$

184. $x + 2 > 1$ **189.** $x + 3 \geq -4$

185. $x - 2 < 9$ **190.** $7 + x \leq 6$

186. $-1 + x < 2$ **191.** $6 + x < 9$

192. $-7 + x \geq -4$

201. $2x > 2$

193. $x + 7 < 4$

202. $3x < -6$

194. $-5 + x \geq -5$

203. $7x > 14$

195. $x - 1 > 1$

204. $12x > 6$

196. $x + 1 \geq 1$

205. $3x \geq 12$

197. $x + 8 < 3$

206. $2x \leq -4$

198. $x - 5 \leq 10$

207. $3x \geq 2$

199. $-2 + x < 7$

208. $3x > 9$

200. $x + 6 > 1$

209. $6x \leq 15$

210. $-2x > 6$

211. $-x \leq 2$

212. $-x < -2$

213. $-6x \geq 15$

214. $2x > -5$

215. $4x \leq 12$

216. $3x > 9$

217. $2x \geq 8$

218. $-x < 2$

219. $-3x \leq -6$

220. $5x \geq -15$

221. $2(x + 1) \geq 4$

222. $5(x - 2) \leq 2$

223. $3(x + 2) \leq 12$

224. $7(2x - 1) > -14$

225. $-2(x - 3) \geq -12$

226. $-(x + 2) > -4$

227. $4(x + 1) < 16$

228. $-3(x + 1) < 4$

237. $4(2x - 2) > (6x + 5)$

229. $8(x - 1) \geq -4$

238. $(6x + 3) > 5(7x - 2)$

230. $5(2x - 3) \leq 25$

239. $-3(2x + 1) \leq 8(x + 4)$

231. $5(x - 1) \leq 5(2x - 3)$

240. $8(8x + 4) \geq 6(3x + 5)$

232. $4(x + 2) < 3(5x + 7)$

241. $3(2x - 6) \leq 2(8x - 3)$

233. $5(3x + 4) \leq 5(2x + 9)$

242. $-5(2x + 3) \geq -10(3x - 2)$

234. $6(10x - 5) \geq 5(6x - 2)$

243. $9(9x - 5) \geq 6(6x + 1)$

235. $-7(7x + 2) \leq 5(4x - 6)$

244. $6(2x - 10) < 7(7x + 6)$

236. $2(15x - 3) < 5(x - 4)$

245. $7(2x + 7) \leq 9(8x + 10)$

246. $4(6x - 6) > -2(7x - 8)$

247. $-9(2x + 9) < 5(2x + 7)$

248. $2(2x - 7) \leq (x - 9)$

249. $3(4x - 3) > 10(x + 2)$

250. $(2x + 8) \geq -5(9x + 8)$

251. $-2 < x + 1 < 4$

252. $-1 > x + 3 > 3$

253. $5 > x + 5 > 12$

254. $-2 < x + 1 < -4$

255. $1 < x + 7 > 4$

256. $5 > x + 7 < 7$

257. $4 < x + 1 < 5$

258. $-18 > x + 9 > 27$

259. $7 < x + 7 < 8$

260. $5 > x + 3 > 2$

261. $|x| > 2$

262. $|x| \leq 3$

263. $|x| < 4$

264. $|x| \geq 5$

265. $|x| \leq 7$

266. $|x| \leq 1$

267. $|x| > 12$

268. $|x| \geq 6$

269. $|x| \geq 8$

270. $|x + 1| < 4$

271. $|x + 2| > 3$

272. $|x - 3| \geq 6$

273. $|x + 4| \leq 2$

274. $|x - 2| \geq 6$

275. $|x - 7| < 5$

276. $|x - 5| > 4$

277. $|x - 7| \geq 12$

278. $|x + 4| \leq 2$

279. $|x + 5| \geq 7$

280. $2|x + 1| \geq 4$

281. $3|x + 2| \leq 3$

282. $2|x - 3| < 6$

283. $4|x + 7| \geq 12$

284. $|x + 8|2 < 8$

285. $6 \geq 3|x - 3|$

286. $2|x + 1| > 1$

287. $6|x + 3| \geq 18$

288. $8 \geq 4|x - 3|$

289. $12 \geq 2|x + 1|$

290. $4|x + 2| \geq 6$

Chapter 5 Linear Equations

Most linear equations are written in y-intercept form, which is y=mx+b. In this form, m and b are constants representing the slope and y-intercept, respectively. The slope is the rise divided by the run (m=rise/run).

To determine the slope and y-intercept of an equation, we must first put the equation in y-intercept form. In order to put an equation in y-intercept form, we must solve for y. Below is an example of this process:

Determine the slope and y-intercept of the following equation:

$$6x + 2y = 12$$

First, isolate the y-term on the left side of the equation:

$$2y = -6x + 12$$

Now divide both sides by 2 isolate the y variable:

$$\frac{2y}{2} = \frac{-6x}{2} + \frac{12}{2}$$

$$y = -3x + 6$$

Now the equation is in y-intercept form. The slope (m) equals -3. The y-intercept is 6.

In other types of problems, we are given two points and asked to find the slope between them.

The following equation is used to determine the slope between two points:

$$slope = \frac{y_2 - y_1}{x_2 - x_1}$$

Find the slope of the line that passes between points (2, 3) and (1, 5).

Here, $x_1=2$, $y_1=3$ and $x_2=1$, $y_2=5$:

$$Slope = \frac{5-3}{1-2} = \frac{2}{-1} = -2$$

Therefore, the slope is -2.

Find the y-intercept (b) of the line with a slope of 3 and that passes through points (4, 5):

Since we already know the slope, we know:

$$y = 3x + b$$

To find b, we substitute points (4, 5) into the equation and solve for b:

$$5 = 3(4) + b$$

$$5 = 12 + b$$

$$5 - 12 = b$$

$$-7 = b$$

Therefore:

$$y = 3x - 7$$

Practice Problems

291. What is the equation for the graph below?

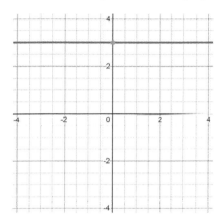

292. What is the equation for the graph below?

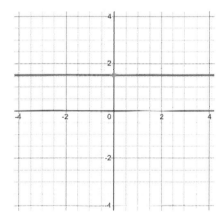

293. What is the equation for the graph below?

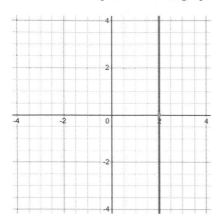

294. What is the equation for the graph below?

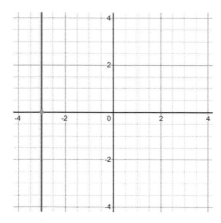

295. What is the equation for the graph below?

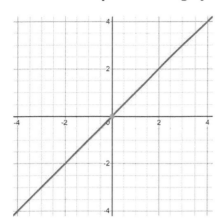

Convert the below equations into y-intercept form:

296. $2y = 4x + 6$

297. $4y = 4x + 12$

298. $6y = 12x + 24$

299. $5y = 15x + 25$

300. $7y = 14x + 28$

301. $10y = 20x + 40$

302. $12y = -48x + 144$

303. $9y = 27x + 45$

304. $11y = -44x + 121$

305. $8y = 24x + 40$

306. $2y = x + 2$

307. $10y = 10x + 5$

308. $-3y = 6x - 5$

309. $12y = -3x + 6$

310. $14y = 7x + 28$

311. $30y = 10x - 15$

312. $-17y = 8x - 18$

313. $25y = -75x + 100$

314. $7y = 28x - 63$

315. $9y = 54x + 81$

316. $-2y + 4x = 12$

317. $4y - 16x = 20$

318. $4y + 16x = 36$

319. $-25y + 5x = 35$

320. $7y = 14x + 28$

321. $11y - 22x = -66$

322. $12y + 4x = -156$

323. $2y + 24x = 4$

324. $-10y + 55x = 135$

325. $8y + 8x = 24$

326. $-4y - x = 2$

327. $9y + 3x = 6$

328. $-6y + 9x = 12$

329. $2y - 4x = 16$

330. $7y - 21x = 49$

331. $15y + 5x = -10$

332. $-7y + 21x = 84$

333. $5y + 15x = -50$

334. $-13y + 39x = 52$

335. $9y + 18x = -51$

336. $3y = 5x - 45$

337. $7y = 3x - 91$

338. $8y = -10x + 16$

339. $-13y = 12x + 26$

340. $-2y = 6x - 2$

341. $-9y = -2x + 72$

342. $8y = -6x - 68$

343. $5y = 7x + 60$

344. $13y = 14x + 169$

345. $-8y = 64x + 28$

346. $12y = 7x + 10$

347. $11y = x - 132$

348. $7y = -x - 7$

349. $-18y = -18x + 28$

350. $-13y = 6x + 14$

351. $3y = -8x - 16$

352. $7y = 21x + 25$

353. $9y = -72x + 81$

354. $-5y = -25x + 55$

355. $-12y = -14x - 60$

356. $15y + 13x = -14$

357. $4y - 16x = -32$

358. $-13y + 26x = 44$

359. $22y - 44x = 33$

360. $-18y = -45x + 36$

361. $14y + 28x = 56$

362. $6y + 12x = 8$

363. $-20y - 20x = -40$

364. $25y + 50x = -125$

365. $-6y - 18x = 16$

366. $16y + 32x = 14$

367. $24y + 48x = 35$

368. $9y + 45x = 34$

369. $-4y + 44x = 24$

370. $-8y + 28x = -98$

371. $7y + 12x = 120$ **374.** $4y - 42x = -4$

372. $7y + 17x = 6$ **375.** $10y - 8x = 7$

373. $-14y - 16x = 12$

What is the slope of a line that connects the two points?

376. $(1, 1)$ and $(5, 1)$ **387.** $(6, 1)$ and $(6, 4)$

377. $(2, 4)$ and $(1, 6)$ **388.** $(9, 8)$ and $(9, 8)$

378. $(5, 6)$ and $(6, 3)$ **389.** $(2, 3)$ and $(0, 0)$

379. $(9, 4)$ and $(8, 2)$ **390.** $(-7, 5)$ and $(8, -1)$

380. $(3, 6)$ and $(10, 7)$ **391.** $(2, 7)$ and $(-6, 6)$

381. $(6, 7)$ and $(8, 10)$ **392.** $(12, -8)$ and $(0, 4)$

382. $(4, 2)$ and $(1, 11)$ **393.** $(10, 2)$ and $(6, -12)$

383. $(1, 8)$ and $(10, 4)$ **394.** $(-7, 10)$ and $(5, 4)$

384. $(3, 7)$ and $(9, 4)$ **395.** $(8, 11)$ and $(4, 6)$

385. $(5, 10)$ and $(2, 7)$ **396.** $(5, -12)$ and $(11, 8)$

386. $(10, 8)$ and $(7, 2)$ **397.** $(9, 1)$ and $(-12, 11)$

398. (-1, 8) and (-8, 4)

400. (7, 12) and (1, -9)

399. (4, -9) and (6, 8)

What is the y-intercept form of the line described below?

401. A line with a slope of 1 that passes through points (3, 5)

402. A line with a slope of 0 that passes through points (8, 6)

403. A line with a slope of 2 that passes through points (-1, 5)

404. A line with a slope of 5 that passes through points (-3, -4)

405. A line with a slope of 4 that passes through points (4, 4)

406. A line with a slope of 8 that passes through points (-1, -1)

407. A line with a slope of -2 that passes through points (5, 5)

408. A line with a slope of 4 that passes through points (6, 7)

409. A line with a slope of -2 that passes through points (9, 1)

410. A line with a slope of 10 that passes through points (-3, -2)

411. A line with a slope of -3 that passes through points (11, 10)

412. A line with a slope of 5 that passes through points (1, 5)

413. A line with a slope of 4 that passes through points (8, 12)

414. A line with a slope of -3 that passes through points (-12, 7)

415. A line with a slope of 7 that passes through points (-5, -4)

416. A line with a slope of 2 that passes through points (-6, 6)

417. A line with a slope of -2 that passes through points (1, -1)

418. A line with a slope of 4 that passes through points (3, 2)

419. A line with a slope of 7 that passes through points (9, 9)

420. A line with a slope of 4 that passes through points (2, 0)

421. A line with a slope of 8 that passes through points (-2, -6)

422. A line with a slope of -4 that passes through points (7, -8)

423. A line with a slope of 2 that passes through points (3, 3)

424. A line with a slope of 3 that passes through points (7, 7)

425. A line with a slope of 9 that passes through points (-8, -7)

426. A line with a slope of 11 that passes through points (0, 0)

427. A line with a slope of 10 that passes through points (0, -2)

428. A line with a slope of -6 that passes through points (-1, 7)

429. A line with a slope of 4 that passes through points (10, 6)

430. A line with a slope of -3 that passes through points (1, 0)

Chapter 6 Systems of Linear Equations

Systems of Linear Equations can be solved either by the substitution method or addition method. Below is an example of the substitution method:

$$3y + 3x = 6$$

$$y = x + 4$$

Since we know y=x+4, we can substitute x+4 into y in the first equation:

$$3(x + 4) + 3x = 6$$

$$3x + 12 + 3x = 6$$

$$6x = 6 - 12 = -6$$

$$\frac{6x}{6} = \frac{-6}{6}$$

$$x = -1$$

Now that we have solved for x, we can plug this value back into either equation and solve for y. Here, we use the second equation:

$$y = x + 4$$

$$y = -1 + 4 = 3$$

Therefore, the solution to this linear equation is $x = -1$, $y = 3$. This is to say, the two linear equations intersect at point (-1, 3).

Solve the two equations below using the addition method:

$$-4x + y = 10$$

$$\underline{4x + 2y = 5}$$

$$3y = 15$$

We can cancel x variable because -4x+4x=0. Now we can solve for y:

$$\frac{3y}{3} = \frac{15}{3}$$

$$y = 5$$

Now we can substitute y=5 back into either equation. (NOTE: Suggest to always pick the easier equation):

$$-4x + y = 10$$

$$-4x + 5 = 10$$

$$-4x = 10 - 5 = 5$$

$$\frac{-4x}{-4} = \frac{5}{-4}$$

$$x = -\frac{5}{4}$$

The solution is therefore, $y = 5, x = -\frac{5}{4}$

Practice Problems

Solve each system of linear equations using substitution method:

431. $2y + 2x = 4$
    $y = x + 2$

437. $6y + 3x = 3$
    $y = 2x + 3$

432. $3y - 3x = 9$
    $y = 2x - 1$

438. $8y - 2x = -3$
    $y = 4x - 6$

433. $4y + 2x = 8$
    $y = x + 1$

439. $9y + 5x = 14$
    $y = 5x - 4$

434. $3y - 7x = 5$
    $y = x - 5$

440. $7y - 2x = 24$
    $y = 6x - 8$

435. $6y - 3x = -9$
    $y = x + 4$

441. $4y - 3x = 1$
    $y = 2x - 6$

436. $7y + 5x = 17$
    $y = 3x - 5$

442. $7y + 8x = 1$
    $y = 6x - 7$

443. $6y + 4x = -5$
$\quad y = 7x + 3$

450. $2y + 4x = 6$
$\quad 2y = 6x - 4$

444. $9y - 3x = -42$
$\quad y = 7x - 8$

451. $2y + 4x = 6$
$\quad 3y - 3x = 9$

445. $5y + 9x = 16$
$\quad y = x - 8$

452. $2y + 4x = 6$
$\quad 4y + 4x = 8$

446. $2y - 2x = 6$
$\quad y = 3x - 3$

453. $4y - 2x = 6$
$\quad 3y + 3x = 9$

447. $8y + 9x = 9$
$\quad y = 5x - 5$

454. $2y + 4x = 20$
$\quad 5y + 2x = 10$

448. $5y + 5x = 5$
$\quad y = 6x - 6$

455. $2y - 5x = 10$
$\quad 3y - 6x = 12$

449. $7y - 6x = -4$
$\quad y = 2x + 4$

456. $4y + 6x = 20$
$\quad 2y + 4x = 8$

457. $4y - 8x = 12$
$2y - 5x = 10$

464. $3y + 3x = 6$
$8y - 16x = -32$

458. $6y + 6x = 18$
$4y + 8x = 8$

465. $5y - 15x = 15$
$3y - 4x = 19$

459. $5y - 5x = 25$
$9y + 6x = 15$

466. $15y + 10x = 5$
$2y - 2x = 4$

460. $7y - 7x = 21$
$6y - 8x = 24$

467. $2y - 6x = 4$
$4y - 10x = 6$

461. $8y + 6x = 28$
$2y + 4x = 12$

468. $7y - 14x = 7$
$8y - 14x = 4$

462. $7y - 8x = 54$
$3y + 6x = 42$

469. $4y + 8x = 0$
$2y - 4x = 8$

463. $6y - 6x = -24$
$7y + 7x = 28$

470. $4y + 4x = 12$
$6y + 4x = 4$

471. $12y + 3x = 6$
$3y + 6x = 12$

478. $4y - 8x = 8$
$14y + 12x = -12$

472. $6y + 3x = 9$
$2y + 6x = 18$

479. $9y - 6x = 18$
$5y - 5x = 25$

473. $14y + 20x = 10$
$3y + 6x = 9$

480. $8y + 16x = 24$
$4y + 6x = 4$

474. $6y - 8x = 12$
$4y - 8x = 16$

481. $5y - 10x = 25$
$4y + 3x = 9$

475. $6y + 9x = 18$
$4y + 8x = 16$

482. $7y - 8x = 4$
$14y - 14x = 7$

476. $3y - 6x = 18$
$3y + 3x = 9$

483. $6y + 4x = 34$
$8y + 40x = -24$

477. $6y + 18x = 12$
$4y + 2x = 18$

484. $9y + 27x = 72$
$6y + 12x = 18$

485. $9y + 36x = 81$
$5y + 5x = 60$

492. $9y - 4x = 13$
$12y - 48x = 60$

486. $4y + 20x = 40$
$9y + 27x = 54$

493. $14y + 6x = 86$
$5y - 40x = -180$

487. $18y + 120x = 768$
$3y + 60x = 48$

494. $23y + 12x = 81$
$17y - 34x = 17$

488. $25y + 50x = 75$
$60y + 20x = -20$

495. $9y + 18x = 18$
$16y + 9x = 9$

489. $14y - 21x = -7$
$5y - 15x = 20$

496. $5y - 15x = 25$
$7y - 6x = 5$

490. $7y + 13x = 17$
$8y - 8x = 8$

497. $6y + 7x = -10$
$9y + 27x = 18$

491. $4y + 16x = 8$
$6y - 12x = 12$

498. $2y + 12x = 16$
$8y - 9x = 7$

499. $5y - 25x = 15$
$3y + 15x = 39$

505. $6y - 12x = 8$
$8y - 12x = 20$

500. $12y + 12x = 12$
$14y + 9x = 9$

506. $20y + 5x = 10$
$20y + 9x = 8$

501. $16y + 8x = 8$
$18y + 2x = 23$

507. $12y + 10x = 56$
$14y - 14x = 14$

502. $14y - 9x = 29$
$10y - 5x = 15$

508. $8y + 4x = 4$
$10y - 6x = -6$

503. $5y + 9x = 53$
$15y - 5x = 15$

509. $6y + 9x = 2$
$4y - 8x = 6$

504. $25y + 12x = 3$
$20y + 10x = 10$

510. $10y + 5x = -10$
$16y - 7x = 14$

Solve each system of linear equations using addition method:

511. $2y + 2x = 4$
$-2y + x = 2$

517. $6y + 3x = 3$
$-6y - 2x = 3$

512. $-3y - 3x = 9$
$3y + 2x = 1$

518. $8y - 2x = -3$
$-8y - 4x = 15$

513. $4y + 2x = 8$
$-4y - x = 1$

519. $9y + 5x = 14$
$-9y + 5x = 6$

514. $3y - 7x = 5$
$-3y + x = 7$

520. $-7y - 2x = 24$
$7y + 6x = -8$

515. $-6y - 3x = -9$
$6y - x = 5$

521. $4y - 3x = 1$
$-4y + 2x = 6$

516. $7y + 5x = 17$
$-7y + 3x = 7$

522. $-7y + 8x = 1$
$7y - 6x = 7$

523. $-6y + 4x = -5$
$\quad 6y + 7x = 16$

524. $9y - 3x = -42$
$\quad -9y - 7x = 2$

525. $-y + 9x = 16$
$\quad y - x = 8$

526. $2y - 2x = 7$
$\quad -2y + 4x = 3$

527. $8y + 9x = 17$
$\quad -8y - 5x = -5$

528. $5y + 5x = 5$
$\quad -5y - 6x = 6$

529. $7y - 6x = -4$
$\quad -7y + 2x = 4$

530. $2y + 4x = 8$
$\quad 2y - 4x = 4$

531. $2y + 3x = 6$
$\quad 3y - 3x = 9$

532. $2y - 4x = 16$
$\quad 4y + 4x = 8$

533. $4y - 3x = 5$
$\quad 3y + 3x = 9$

534. $2y - 2x = 20$
$\quad 8y + 2x = 10$

535. $-2y - 5x = 10$
$\quad 3y + 5x = -4$

536. $4y - 4x = -20$
$\quad -2y + 4x = 8$

537. $-4y - 8x = 12$
$2y + 8x = -10$

544. $3y - 9x = -6$
$8y - 3x = -44$

538. $6y + 6x = 18$
$-4y - 6x = -8$

545. $-5y - 15x = 17$
$-3y + 5x = 13$

539. $-5y - 5x = 25$
$10y + 5x = -15$

546. $15y + 10x = 5$
$2y - 2x = 4$

540. $7y - 7x = 21$
$-6y + 7x = -24$

547. $-3y - 5x = 4$
$-4y - 10x = 6$

541. $8y + 6x = 24$
$-2y + 4x = 16$

548. $7y - 14x = 19$
$-8y - 14x = 4$

542. $-7y - 3x = 54$
$3y + 6x = 46$

549. $4y + 8x = 0$
$2y - 4x = 8$

543. $6y - 3x = -24$
$7y - 3x = 28$

550. $-4y + 4x = 15$
$6y + 8x = 16$

551. $12y + 3x = 9$
$\quad\ 3y + 6x = -3$

552. $6y + 18x = 9$
$\quad\ 4y + 6x = -3$

553. $-12y + 20x = -8$
$\quad\ \ -3y + 6x = -2$

554. $6y - 8x = 12$
$\quad\ 4y - 8x = 16$

555. $-8y + 9x = 9$
$\quad\ \ 4y + 8x = 8$

556. $-3y - 6x = 18$
$\quad\ \ -3y + 3x = 9$

557. $6y + 18x = 18$
$\quad\ 4y + 2x = 12$

558. $4y - 8x = 56$
$\quad\ \ 16y + 12x = 4$

559. $-10y - 9x = -21$
$\quad\ \ 5y - 5x = 20$

560. $8y + 16x = 20$
$\quad\ \ -4y + 6x = 4$

561. $5y - 10x = 29$
$\quad\ 5y + 3x = 3$

562. $-7y - 8x = 5$
$\quad\ \ -14y - 14x = 8$

563. $-6y + 4x = 3$
$\quad\ \ 8y + 40x = -38$

564. $-9y + 27x = 60$
$\quad\ \ 6y + 12x = 20$

565. $9y + 36x = 81$
$\quad 5y + 6x = 10$

572. $-9y - 4x = 8$
$\quad -12y - 8x = 10$

566. $-2y + 6x = 2$
$\quad 3y + 3x = 9$

573. $10y + 6x = 17$
$\quad 5y - 4x = 12$

567. $18y + 120x = 24$
$\quad 3y + 60x = 48$

574. $23y + 12x = 27$
$\quad 16y - 36x = 4$

568. $120y + 60x = 80$
$\quad 60y + 20x = -20$

575. $9y + 18x = 23$
$\quad -16y + 9x = -9$

569. $-14y - 21x = -7$
$\quad -5y - 7x = 20$

576. $5y - 15x = 31$
$\quad 7y - 5x = 5$

570. $-8y + 13x = 17$
$\quad 8y - 8x = 8$

577. $-6y + 7x = 10$
$\quad 9y + 6x = 18$

571. $4y + 6x = 30$
$\quad 6y - 2x = 12$

578. $-2y + 12x = 16$
$\quad 8y - 9x = 14$

579. $5y - 25x = 5$
$3y + 15x = 33$

585. $6y - 12x = 8$
$3y - 12x = 20$

580. $2y + 12x = 12$
$2y + 9x = 9$

586. $-20y + 5x = 10$
$20y + 9x = 4$

581. $9y + 8x = 8$
$18y + 2x = 30$

587. $4y + 10x = 14$
$16y - 14x = 2$

582. $15y - 9x = 6$
$10y - 5x = 15$

588. $8y + 4x = 9$
$-10y - 8x = -6$

583. $-5y + 9x = 12$
$15y - 5x = 30$

589. $6y + 9x = 2$
$-4y - 8x = 6$

584. $-25y + 20x = 10$
$20y + 4x = 12$

590. $10y + 5x = -10$
$16y - 7x = 14$

Chapter 7 Polynomials

A polynomial is an equation with multiple terms, including variables and constants. Below are some examples of polynomials.

Solve the following:

$$x + x = ?$$

The answer is:

$$x + x = 2x$$

Solve the following:

$$x^2 + x^2 = ?$$

The answer is:

$$x^2 + x^2 = 2x^2$$

Solve the following:

$$x^2 + x = ?$$

The answer is:

$$x^2 + x = x^2 + x$$

x^2 and x and cannot be added because they are not like terms. The numbers 1 and 2 are like terms, x and x are like terms, x^2 and x^2 are like terms, but x and x^2 are not like terms. Therefore they cannot be added together.

Solve the following:

$$(x + x^2) + (x + x^2) = ?$$

The answer is:

$$(x + x^2) + (x + x^2) = 2x + 2x^2$$

Solve the following:

$$(1 + x + x^2) + (2 + x + x^2) =?$$

The answer is:

$$3 + 2x + 2x^2$$

The following problems are more complicated. But the rules are the same, combine like terms and add. If it is a subtraction problem, then first change the signs and then add:

$$(2y^2 + 3xy + 4x) - (y^2 + 2xy - 2x^2)$$

Here, we have to change the sign of each term:

$$(2y^2 + 3xy + 4x) + (-y^2 - 2xy + 2x^2)$$
$$y^2 + xy + 4x + 2x^2$$

Solve below equation:

$$x \cdot x =?$$

The answer is:

$$x \cdot x = x^2$$

Below is another example:

$$5x^4 \cdot 5x^3 = 5 \cdot 5x^{4+3} = 25x^7$$

In the above problems, the variable was always x. Below, we introduce multiple variables. In this case, the exponents for each variable must be added separately:

$$(x^3y^2)(x^2y^3) = x^{3+2}y^{2+3} = x^5y^5$$

In some problems, there are not always the same variables to multiply:

$$(x^3y^2z)(x^2y^3) = x^{3+2}y^{2+3}z = x^5y^5z$$

Below is another layer of complexity added to the same problem:

$$(2x^3y^2z)(3x^2y^3) = 2 \cdot 3x^{3+2}y^{2+3}z = 6x^5y^5z$$

Similarly, we can divide polynomials by monomials:

$$\frac{6x^3 - 9x^2}{3x}$$

$$\frac{6x^3}{3x} - \frac{9x^2}{3x}$$

$$\frac{6}{3}x^{3-1} - \frac{9}{3}x^{2-1}$$

$$2x^2 - 3x$$

Practice Problems

Add the below polynomials:

591. $(3 + 3x) + (1 + 2x)$

592. $(4 + 2x) + (2 + 5x)$

593. $(7 + 2x) + (1 + 7x)$

594. $(1 + 9x) + (8 + 4x)$

595. $(9 + 2x) + (3 + 3x)$

596. $(12 + 8x) + (9 + 2x)$

597. $(5 + 12x) + (4 + 12x)$

598. $(8 + 7x) + (7 + 13x)$

599. $(11 + 6x) + (2 + 6x)$

600. $(2 + x) + (11 + 9x)$

601. $(6 + 5x) + (5 + 2x)$

602. $(1 + 9x) + (5 + 6x)$

603. $(8 + 6x) + (10 + 8x)$

604. $(7 + 2x) + (6 + 3x)$

605. $(3 + 6x + 3x^2) + (14 + 15x + x^2)$

606. $(12 + 4x + 10x^2) + (3 + 3x + 4x^2)$

607. $(1 + 5x + 12x^2) + (18 + 6x + 13x^2)$

608. $(6 + 12x + 12x^2) + (11 + 4x + 15x^2)$

609. $(10 + 6x + 9x^2) + (4 + 11x + 10x^2)$

610. $(3 + 11x + 4x^2) + (5 + x + 8x^2)$

611. $(9 + x + 6x^2) + (21 + 7x + 5x^2)$

612. $(5 + 18x + 10x^2) + (5 + 6x + 12x^2)$

613. $(7 + 17x + 7x^2) + (17 + 12x + 12x^2)$

614. $(8 + 7x + 9x^2) + (6 + 17x + 7x^2)$

615. $(9 + 13x + 8x^2) + (7 + 8x + 10x^2)$

616. $(12 + 15x - 8x^2) + (12 + 4x + 9x^2)$

617. $(6 - 24x + 3x^2) + (4 + 9x + 2x^2)$

618. $(5 + 4x - 12x^2) + (9 - 10x + 3x^2)$

619. $(11 + 9x + 6x^2) + (9 + 13x + 4x^2)$

620. $(15 - 5x - 11x^2) + (7 - 8x - 10x^2)$

621. $(4 + 8x - 10x^2) + (10 - 2x + 6x^2)$

622. $(3 + 6x + 7x^2) + (6 + 6x + 11x^2)$

623. $(4 + 3x - 4x^2) + (12 - 4x + x^2)$

624. $(2 - 12x + 5x^2) + (8 + 12x + 7x^2)$

625. $(18 + 20x - 4x^2) + (3 + 3x + 13x^2)$

Subtract the below polynomials:

626. $(3 + 6x + 3x^2) - (14 + 15x + x^2)$

627. $(12 + 4x + 10x^2) - (3 + 3x + 4x^2)$

628. $(1 + 5x + 12x^2) - (18 + 6x + 13x^2)$

629. $(6 + 12x + 12x^2) - (11 + 4x + 15x^2)$

630. $(10 + 6x + 9x^2) - (4 + 11x + 10x^2)$

631. $(3 + 11x + 4x^2) - (5 + x + 8x^2)$

632. $(9 + x + 6x^2) - (21 + 7x + 5x^2)$

633. $(5 + 18x + 10x^2) - (5 + 6x + 12x^2)$

634. $(7 + 17x + 7x^2) - (17 + 12x + 12x^2)$

635. $(8 + 7x + 9x^2) - (6 + 17x + 7x^2)$

636. $(9 + 13x + 8x^2) - (7 + 8x + 10x^2)$

637. $(12 + 15x - 8x^2) - (12 + 4x + 9x^2)$

638. $(6 - 24x + 3x^2) - (4 + 9x + 2x^2)$

639. $(5 + 4x - 12x^2) - (9 - 10x + 3x^2)$

640. $(11 + 9x + 6x^2) - (9 + 13x + 4x^2)$

641. $(15 - 5x - 11x^2) - (7 - 8x - 10x^2)$

642. $(4 + 8x - 10x^2) - (10 - 2x + 6x^2)$

643. $(3 + 6x + 7x^2) - (6 + 6x + 11x^2)$

644. $(4 + 3x - 3x^2) - (12 - 4x + x^2)$

645. $(2 - 12x + 5x^2) - (8 + 12x + 7x^2)$

646. $(18 + 20x - 4x^2) - (3 + 3x + 13x^2)$

Simplify the below expression:

647. $(2x^3 3yz^2)(2x^2 3y^3 z^3)(x^2 y^3 4z^5)$

648. $(3x^2y^22z^3)^3(2x^5y^4z^5)(x^72y^32z^4)^2$

649. $(x^44y^33z^4)(x^4y^5z^2)(3x^62y^8z^2)^3$

650. $(x2y^44z^5)^2(x^43y^2z)^3(2x^2y^3z)^2$

651. $(x^4y^57z^6)(4x^7y^7z^4)(2x^2y^5z^2)^2$

652. $(x^32y^{12}z^4)(x^92y^9z^6)(x^33y^83z^6)$

653. $(x^4y^6z^7)^3(x^24y^4z^7)(x^25y^2z^4)^2$

654. $(xy^24z^2)(x^92y^3z^5)^4(x^55y^3z^1)$

655. $(x^6y^32z^3)(x^52y^6z^2)(6x^62z^2)$

656. $(x^25y^23z^6)(x^44y^7z)(x^4y^4z^4)^2$

657. $(4x^4y^75z^4)(x^64y^{11}z^5)(x^2y^2z^4)^2$

658. $(x^7y^4z^2)(x^7y^32z^6)^3(2x^3y^75z^8)^2$

659. $(2x^2y^5z^0)^4(x^24y^4z^8)(4x^{11}y^34z^6)$

660. $(2x^42y3z^8)(x^46y^5z^7)(x^8y^32z^2)^2$

661. $(2x^5y^210z^4)(2x^3y^5z^2)(x^6y^43z^8)$

662. $(x^25y^7z^5)(8x^8z^4)(x^44y^6z^7)$

663. $(10x^24y^3z^2)(x^2y^3z)(x^33y^9z^{12})$

664. $(3x^2yz^5)^2(x^32y^2z^3)^2(x2y^{14}2z^8)^2$

665. $(x^46y^33z^8)(x^74yz^2)(x^7y^34z^4)$

666. $(x^5y^64z)(x^53y^3z)(2x^2y^23z^6)^3$

667. $(9x^2yz^4)(x^64y^6z^5)(x^4y^42z^8)^2$

668. $(x^4yz^2)(6x^2y^4z^3)(4x^3yz^2)$

669. $(2x^3y^5z^2)(x^67y^3z^5)^5(x^24y^{12}z^4)$

670. $(x4yz^2)(3x^6y^2z^6)(x^63y^4z^5)^4$

671. $(x^23y^5z^2)^4(x4y^8z^7)^3(3x^5y^5z^7)^2$

672. $(x^22y^3z^2)^7(2x^3y^3z^8)^6(x^2y^22z^6)^3$

673. $\dfrac{(2x^2 3y^3 z^4)(8x^5 6y^3 z^2)}{(6x^2 y^2 3z^3)}$

674. $\dfrac{(6x^3 5y^4 z^6)(4x^4 3y^8 z^2)}{(3x^3 2y^7 8z)}$

675. $\dfrac{(x^3 6y^3 5z^2)(4x^4 y^3 3z^6)}{(x^8 8y^7 2z^3)}$

676. $\dfrac{(5x^8 6y^9 4z^6)(x^5 y^6 z^6)}{(2x^2 y^3 3z^2)}$

677. $\dfrac{(5x^4 3y^2 2z^7)(6x^3 4y^8 z^6)}{(8x^2 2y^3 3z^2)}$

678. $\dfrac{(x^2 4y^6 2z^8)(5x^7 6y^4 z^2)^2}{(2x^6 2y^4 3z^3)}$

679. $\dfrac{(x^3 y^4 3z^2)(5x^6 6y^4 8z^8)}{(4x^4 6y^6 z^9)}$

680. $\dfrac{(x^2 6y^2 5z^3)(3x^3 2y^4 z^5)}{(2x^7 4y^8 2z^7)}$

681. $\dfrac{(6x^9 5y^4 z^8)(x^3 y^6 z^2)}{(3x^2 4y^3 2z^5)}$

682. $\dfrac{(6x^6 3y^2 5z^4)(x^3 2y^7 z^8)}{(2x^3 y^4 z^5)}$

683. $\dfrac{(x^4 4y^6 6z^2)(5x^4 3y^3 2z^6)}{(6x^7 2y^2 z^8)}$

684. $\dfrac{(3x^2 y^7 2z^5)(x^6 y^9 4z^4)}{(3x^3 4y^2 4z^7)}$

685. $\dfrac{(8x^2 y^7 2z^5)(4x^3 y^9 5z^6)}{(2x^8 y^2 4z^4)}$

686. $\dfrac{(x^6 5y^5 4z^4)(2x^2 3y^4 z^5)}{(2x^2 y6z^2)}$

687. $\dfrac{(6x^2 5y^3 z^6)(4x^2 y^4 z^8)}{(x^7 2y^5 3z^9)}$

688. $\dfrac{(3x^2 2y^6 z^4)(x^5 5y^7 z)}{(8x^8 7y^3 z^9)}$

689. $\dfrac{(7x5y^4z^2)(3x^5y^87z^6)}{(7x^7y^6z^4)^2}$

697. $\dfrac{(x^63y^4z^5)(2x^64y^9z^4)}{(2x^7y^46z^8)}$

690. $\dfrac{(x^95y^82z^2)(5x^33y^5z^4)}{(x^6y^4z^7)}$

698. $\dfrac{(x^8y^43z)(4x^3y^5z^8)}{(2x^78y^6z^9)}$

691. $\dfrac{(x^54y^45z^2)(3x^3y^8z^6)}{(2xy^7z^3)}$

699. $\dfrac{(4x^72y^36z^5)(x^22y^3z^6)}{(8x^63y^4z^8)}$

692. $\dfrac{(x^28y^86z^3)(5x^54y^4z^6)}{(2x^94y^62z^7)}$

700. $\dfrac{(x^6y^5z^2)(72x^9y^3z^8)}{(2x^43y^72z^2)^2}$

693. $\dfrac{(x^73y^4z^3)(7x^84y^6z^5)}{(5x^82y^36z^2)}$

701. $\dfrac{(2x^6y^2z^5)(4x^7y^32z)}{(6x^43y^9z^8)}$

694. $\dfrac{(8x^9y^75z^4)^2(x^82y^34z^6)}{(x^28y^9z^4)}$

702. $\dfrac{(x^98y^6z^3)(3x^52y^54z^9)}{(8x^8y^4z^3)}$

695. $\dfrac{(x^75y^52z^6)(3x^34y^9z^9)}{(x^4y8z^8)}$

703. $\dfrac{(x^8y^56z^2)(4x^3y^93z^7)}{(x^4y^66z^2)}$

696. $\dfrac{(8x^8y^25z^5)(3x^62y^74z^9)}{(x^3y^3z^4)}$

704. $\dfrac{(x^92y^3z^6)(3x^5y^44z^7)}{(x^2y6z^4)}$

705. $\dfrac{(x^4y^3z^6)(2x^2y^54z^5)}{(x^36y^7z^8)}$

713. $\dfrac{(3x^8y^32z^6)(7x^54y^2z^7)}{(x^46y^8z^3)}$

706. $\dfrac{(7x8y^4z^5)(4x^63y^9z^2)}{(6x^7y^62z^8)}$

714. $\dfrac{(6x^23y^74z^4)(2x^6y^5z^3)}{(x^88y^2z^3)}$

707. $\dfrac{(8x^23y^92z^6)(7x^52y^3z^3)}{(x^8y^46z^7)}$

715. $\dfrac{(7x^34y^92z^5)(x^3y^73z^5)}{(x^28y^4z^6)}$

708. $\dfrac{(x^34y^5z^2)(3x2y^78z^5)}{(x^66y^9z^4)}$

716. $\dfrac{(6x^8y^4z^4)(2x4y^2z^4)}{(3x^6y^58z^7)}$

709. $\dfrac{(x^5y^2z^7)(4x^37y^5z^8)}{(3x^42y^4z^9)^3}$

717. $\dfrac{(x^3y^22z^9)(7x^54y^3z^7)}{(3x^6y^46z^8)}$

710. $\dfrac{(8x^93y^57z^3)(4x^7yz^3)}{(2x^46y^6z^6)}$

718. $\dfrac{(x^72y^24z^8)(x^37y^2z^6)}{(2x^47y^5z)}$

711. $\dfrac{(x^83y^84z^3)(x^3y^5z^2)}{(6x^2y^42z^6)}$

719. $\dfrac{(x^23y^66z^4)(7x^92y^3z^3)}{(4x^52y^4z^7)^2}$

712. $\dfrac{(x^76y^4z^5)(3x^64y^47z^9)}{(8x^5y^9z)}$

720. $\dfrac{(4x^32y^27z^4)(4x^73y^8z^6)}{(4x^5y^22z^3)}$

721. $\dfrac{(5x^42y^23z^4)+(x^94y^34z^8)}{(2x^23y^6z^2)}$

722. $\dfrac{(5x^4y^63z^2)+(4x^410y^4z^2)}{(5x^28y^3z^8)}$

723. $\dfrac{(2x^44y^4z^3)+(6x^25y^5z^7)}{(5x^52y^2z^5)}$

724. $\dfrac{(5x^9y^47z^4)+(4x^2y^65z)}{(5x^2y^44z^3)}$

725. $\dfrac{(5x8y^4z^2)+(x^64y^62z^8)}{(10x^7y^22z^4)}$

726. $\dfrac{(6x^85y^54z^4)+(3x^3y^2z^6)}{(2x^2y^46z^9)}$

727. $\dfrac{(x^74y^3z^4)+(3x^66y2z^3)}{(x^55y^26z^3)}$

728. $\dfrac{(7x^82y^5z^4)+(3x^44y^2z^2)}{(6x^7y^32z^6)}$

729. $\dfrac{(x^47y^56z^2)+(2x3y^84z^2)}{(2x^2y^36z^6)^2}$

730. $\dfrac{(2x^8y^27z^2)+(4x^36y^6z^7)}{(2x^93y^3z^5)}$

731. $\dfrac{(4x^3y^65z^2)+(2x^2y2z^9)}{(3x^5y^76z^2)}$

Multiply the two terms below:

732. $(x + 1)(x + 2)$

733. $(x + 2)(x + 7)$

734. $(x - 5)(x + 3)$

735. $(x + 8)(x - 4)$

736. $(x - 6)(x - 2)$

737. $(x + 2)(x + 3)$

738. $(x + 7)(x + 1)$

739. $(x - 3)(x - 5)$

740. $(x + 8)(x + 4)$

741. $(x - 9)(x + 6)$

742. $(x + 8)(x + 2)$

743. $(x + 4)(x + 8)$

744. $(x + 2)(x - 8)$

745. $(x + 5)(x + 7)$

746. $(x - 2)(x + 8)$

747. $(x + 6)(x + 5)$

748. $(x - 3)(x + 5)$

749. $(x - 1)(x + 3)$

750. $(x + 1)(x + 9)$

751. $(x - 7)(x - 7)$

752. $(x + 8)(x + 4)$

753. $(x + 6)(x + 3)$

754. $(x + 2)(x + 4)$

755. $(x - 5)(x + 6)$

756. $(x - 9)(x - 2)$

757. $(x + 5)(x + 12)$

758. $(x + 6)(x + 3)$

759. $(x - 3)(x - 4)$

760. $(x + 5)(x - 2)$

761. $(x + 4)(x - 11)$

762. $(x + 10)(x + 1)$

763. $(x - 5)(x + 9)$

764. $(x + 3)(x + 3)$

765. $(x - 8)(x + 9)$

766. $(x + 8)(x - 7)$

767. $(x + 2)(x - 4)$

768. $(x + 9)(x + 8)$

769. $(x - 2)(x + 3)$

770. $(x + 6)(x - 5)$

771. $(x + 3)(x + 7)$

772. $(x - 3)(x + 2)$

773. $(x + 5)(x + 9)$

774. $(x + 3)(x - 2)$

775. $(x + 4)(x + 4)$

776. $(x - 2)(4x + 3)$

777. $(2x + 2)(x - 5)$

778. $(x - 1)(2x + 3)$

779. $(x - 8)(3x + 2)$

780. $(5x + 2)(x - 4)$

781. $(x + 5)(4x + 4)$

782. $(x + 3)(x + 9)$

783. $(2x - 7)(2x + 6)$

Solve for x:

784. $\sqrt{8x + 9} = 5$

785. $\sqrt{12x + 9} = 9$

786. $\sqrt{x + 9} = 7$

787. $\sqrt{15x + 9} = 12$

788. $\sqrt{5x - 14} = 6$

789. $\sqrt{5x + 6} = 9$

790. $\sqrt{9x + 10} = 1$

791. $\sqrt{4x - 19} = 11$

792. $\sqrt{2x + 16} = 12$

793. $\sqrt{4x + 13} = 5$

794. $\sqrt{6x} = 12$

795. $\sqrt{4x + 13} = 1$

796. $\sqrt{8x + 9} = 13$

797. $\sqrt{16x + 20} = 10$

798. $\sqrt{10x + 9} = 7$

799. $\sqrt{x + 4} = 2$

800. $\sqrt{3x + 1} = 8$

Chapter 8 Quadratic Equations

In the equation below, we multiply two polynomials:

$$(x + 2)(x + 3)$$

$$x^2 + 2x + 3x + 6$$

$$x^2 + 5x + 6$$

Factoring is the reverse of this process. When factoring, we take the final expression and work backwards:

$$x^2 + 5x + 6$$

$$(x+?\,)(x+?\,)$$

We want to know the values for "?" that will make the above equation true. What do we know about the two numbers that will fill the question marks?

1. When added together, the two numbers must equal 5. This gives us the following possible pairs: 1+4, 2+3 (zero is always excluded)
2. When multiplied together, the two numbers must equal 6. This gives us the following possible pairs: 1x6, 2x3

Only pair 2 and 3 satisfy both requirements. Therefore, one of the numbers must be 2 and the other must be 3 (order does not matter). Therefore:

$$(x + 2)(x + 3)$$

Practice Problems

Factor the below expressions:

801. $x^2 + 2x + 1$　　　　　　**803.** $x^2 + 4x + 3$

802. $x^2 + 3x + 2$　　　　　　**804.** $x^2 + 4x + 4$

805. $x^2 + 5x + 6$

806. $x^2 + 7x + 12$

807. $x^2 + 6x + 5$

808. $x^2 + 10x + 25$

809. $x^2 + 7x + 12$

810. $x^2 + 9x + 14$

811. $x^2 + 8x + 16$

812. $x^2 + 7x + 12$

813. $x^2 - 1$

814. $x^2 + x - 2$

815. $x^2 - 4$

816. $x^2 - 2x - 3$

817. $x^2 + 2x - 8$

818. $x^2 + x - 2$

819. $x^2 + x - 6$

820. $x^2 - 6x + 9$

821. $x^2 + 6x + 5$

822. $x^2 - 2x + 1$

823. $x^2 + 9x + 18$

824. $x^2 + 11x + 28$

825. $x^2 + 3x - 10$

826. $x^2 + 8x + 7$

827. $x^2 - 3x - 54$

828. $x^2 - 4x - 45$

829. $x^2 + 10x + 9$

830. $x^2 - x - 56$

831. $x^2 + 12x + 27$

832. $x^2 + 6x - 40$

833. $x^2 + 12x + 32$

834. $x^2 + 14x + 45$

835. $x^2 + 2x - 8$

836. $x^2 - 9$

837. $x^2 + 8x + 7$

838. $x^2 + 10x + 16$

839. $x^2 + 2x - 35$

840. $x^2 + 5x - 24$

841. $x^2 + 11x + 18$

842. $x^2 + 12x + 20$

843. $x^2 + 4x - 32$

844. $x^2 + 12x + 27$

845. $x^2 + 10x + 21$

846. $x^2 + 16x + 64$

847. $x^2 - 7x - 30$

848. $x^2 + 11x + 28$

849. $x^2 + 9x + 14$

850. $x^2 + 15x + 50$

851. $x^2 - 18x + 80$

852. $x^2 + 10x + 9$

853. $x^2 - 14x + 49$

854. $x^2 + 18x + 81$

855. $x^2 + 17x + 60$

856. $x^2 - 5x + 6$

857. $x^2 + 9x + 8$

858. $x^2 + x - 56$

859. $x^2 + 10x + 9$

860. $x^2 + 3x - 10$

861. $x^2 + 8x + 12$

862. $x^2 + 5x + 6$

863. $x^2 + 12x + 35$

864. $x^2 + 7x - 18$

865. $x^2 - x - 12$

866. $x^2 - 3x - 28$

867. $x^2 - 36$

868. $x^2 + 10x + 16$

869. $x^2 - 3x - 10$

870. $x^2 - 2x - 15$

871. $x^2 - 16$

872. $2x^2 + x - 1$

873. $2x^2 - 2$

874. $3x^2 + x - 2$

875. $2x^2 + 6x + 4$

876. $x^2 - 9$

877. $4x^2 - 13x - 12$

878. $9x^2 - 9$

879. $2x^2 - 8x - 10$

880. $3x^2 - 3$

881. $4x^2 + 14x + 6$

882. $2x^2 + 4x - 16$

883. $x^2 + 9x + 18$

884. $3x^2 + 28x + 49$

885. $3x^2 + 4x + 1$

886. $x^2 - 3x - 28$

887. $2x^2 - x - 15$

888. $5x^2 - 4x - 1$

889. $9x^2 + 42x + 49$

890. $4x^2 - 4$

891. $x^2 + 12x + 27$

892. $8x^2 - 44x + 36$

893. $2x^2 + 20x + 18$

894. $x^2 - 6x - 40$

895. $10x^2 + 25x + 15$

896. $3x^2 + 19x - 14$

897. $x^2 + 13x + 36$

898. $x^2 + x - 42$

899. $3x^2 + 15x + 12$

900. $x^2 + 2x - 35$

901. $3x^2 + 5x + 2$

902. $x^2 + 6x - 16$

903. $9x^2 + 12x + 4$

904. $4x^2 + 12x + 9$

905. $10x^2 - 29x + 10$

Solve for x using factoring:

906. $x^2 + 4x + 4$

907. $x^2 + 5x + 6$

908. $x^2 + 11x + 30$

909. $x^2 + 8x + 12$

910. $x^2 + 11x + 24$

911. $x^2 + 9x + 8$

912. $x^2 - x - 12$

913. $x^2 + x - 20$

914. $x^2 - x - 90$

915. $x^2 - x - 110$

916. $x^2 - x - 30$

917. $x^2 - 14x + 40$

918. $x^2 + 22x + 120$

919. $x^2 + 15x + 54$

920. $x^2 - 5x - 24$

921. $x^2 - 12x + 35$

922. $x^2 - 2x - 8$

923. $x^2 + 7x - 8$

924. $x^2 + x - 110$

925. $x^2 + 10x + 21$

926. $x^2 + 3x - 40$

927. $x^2 + 9x - 36$

928. $x^2 - 7x - 60$

929. $x^2 + 20x + 96$

930. $x^2 + 8x - 33$

931. $x^2 - 14x + 33$

932. $x^2 + 5x - 84$

933. $x^2 + 16x + 48$

934. $x^2 + 10x - 75$

935. $x^2 - 14x + 48$

936. $x^2 + x - 6$

937. $x^2 - 2x - 48$

938. $x^2 - x - 12$

939. $x^2 - x - 20$

940. $x^2 - x - 12$

941. $x^2 + x - 30$

942. $x^2 - 6x - 72$

943. $x^2 - 12x + 35$

944. $x^2 + 11x + 10$

945. $x^2 + 2x - 168$

946. $x^2 - x - 132$

947. $x^2 - 11x + 30$

948. $x^2 + 3x - 10$

949. $x^2 - 3x - 180$

950. $x^2 + x - 6$

951. $3x^2 + 4x - 4$

952. $4x^2 + 21x + 5$

953. $2x^2 + 8x - 10$

962. $x^2 - 3x - 10$

954. $4x^2 - 4$

963. $x^2 + 11x + 18$

955. $4x^2 + 13x + 3$

964. $x^2 - 16x + 48$

956. $2x^2 + 19x + 35$

965. $10x^2 + 18x + 8$

957. $x^2 + 9x + 20$

966. $x^2 + x - 132$

958. $6x^2 + 34x + 48$

967. $4x^2 + 10x + 6$

959. $4x^2 + 8x + 4$

968. $x^2 - 25$

960. $6x^2 - 19x + 15$

969. $6x^2 + 23x + 7$

961. $3x^2 + 20x + 25$

970. $6x^2 - 14x + 8$

971. $x^2 = 25$

972. $x^2 = 36$

973. $x^2 = 100$

974. $x^2 = 64$

975. $x^2 = 121$

976. $4x^2 = 64$

977. $3x^2 - 5 = 7$

978. $10x^2 = 1{,}000$

979. $x^2 = 10{,}000$

980. $5x^2 - 20 = 480$

981. $2x^2 + 25 = 123$

982. $3x^2 - 13 = 350$

983. $x^2 - 21 = 123$

984. $3x^2 - 30 = -3$

985. $3x^2 + 45 = 192$

986. $9x^2 - 27 = 54$

987. $2x^2 - 60 = 140$

988. $3x^2 + 16 = 124$

989. $8x^2 - 23 = 9$

990. $x^2 - 45 = 211$

Solve the below problems using the Quadratic Equation:

991. $3x^2 + 15x + 18$

992. $x^2 + 4x + 4$

993. $2x^2 + 19x + 35$

994. $5x^2 + 36x + 36$

995. $4x^2 + 25x + 25$

996. $6x^2 + 43x + 42$

997. $2x^2 + 11x + 12$

998. $3x^2 + 14x + 8$

999. $x^2 + 8x + 15$

1000. $4x^2 + 17x + 18$

Problem 1,001

Why do we have to study Algebra, and how does Algebra help us in life?

Answers to Chapter 1 Review

1. $3^5 = 3 \times 3 \times 3 \times 3 \times 3 = 243$

2. $2^3 = 2 \times 2 \times 2 = 8$

3. $-7^5 = -7 \times -7 \times -7 \times -7 \times -7 = -16,807$

4. $-4^3 = -4 \times -4 \times -4 = -64$

5. $7^7 = 7 \times 7 \times 7 \times 7 \times 7 \times 7 \times 7 = 823,543$

6. $2^3 \times 2^2 = 2^{3+2} = 2^5 = 2 \times 2 \times 2 \times 2 \times 2 = 32$

7. $6^2 \times 6^3 = 6^{2+3} = 6^5 = 6 \times 6 \times 6 \times 6 \times 6 = 7,776$

8. $4^4 \times 4 = 4 \times 4 \times 4 \times 4 \times 4 = 1,024$

9. $\frac{5^2}{5^1} = 5^{2-1} = 5$

10. $\frac{4^7}{4^4} = 4^{7-4} = 4^3 = 4 \times 4 \times 4 = 64$

11. $\frac{5^7}{4^4} = \frac{5 \times 5 \times 5 \times 5 \times 5 \times 5 \times 5}{4 \times 4 \times 4 \times 4} = \frac{78,125}{256} = 305\frac{45}{256}$

12. $\frac{5^4}{4^7} = \frac{5 \times 5 \times 5 \times 5}{4 \times 4 \times 4 \times 4 \times 4 \times 4 \times 4} = \frac{625}{16,384} = 0.038$

13. $\frac{77^7}{77^5} = 77^{7-5} = 77^2 = 77 \times 77 = 5,929$

14. $\frac{467^{98}}{467^{97}} = 467^{98-97} = 467^1 = 467$

15. $\left(\frac{6}{7}\right)^5 = \frac{6^5}{7^5} = \frac{6 \times 6 \times 6 \times 6 \times 6}{7 \times 7 \times 7 \times 7 \times 7} = \frac{7,776}{16,807} = 0.46$

16. $\left(\frac{7}{8}\right)^4 = \frac{7^4}{8^4} = \frac{7 \times 7 \times 7 \times 7}{8 \times 8 \times 8 \times 8} = \frac{2,401}{4,096} = 0.59$

17. $\left(\frac{1}{2}\right)^2 = \frac{1^2}{2^2} = \frac{1 \times 1}{2 \times 2} = \frac{1}{4}$

18. $\left(\frac{3}{4}\right)^5 = \frac{3^5}{4^5} = \frac{3 \times 3 \times 3 \times 3 \times 3}{4 \times 4 \times 4 \times 4 \times 4} = \frac{243}{1,024} = 0.24$

19. $(8^2)^4 = 8^{2 \times 4} = 8^8 = 8 \times 8 \times 8 \times 8 \times 8 \times 8 \times 8 \times 8 = 16,777,216$

20. $(6^3)^2 = 6^{3 \times 2} = 6^6 = 6 \times 6 \times 6 \times 6 \times 6 \times 6 = 46,656$

21. $(8)^4 = 8 \times 8 \times 8 \times 8 = 4,096$

22. $(8^4)^2 = 8^{4 \times 2} = 8^8 = 8 \times 8 \times 8 \times 8 \times 8 \times 8 \times 8 \times 8$ $16,777,216$

23. $(2^2)^{-2} = 2^{2 \times -2} = 2^{-4} = \frac{1}{16} = .0625$

24. $2^{-1} = \frac{1}{2} = 0.5$

25. $6^{-2} = \frac{1}{6^2} = \frac{1}{6 \times 6} = \frac{1}{36} = 0.28$

26. $3^{-2} = \frac{1}{3^2} = \frac{1}{3 \times 3} = \frac{1}{9} = 0.11$

27. $(8^2)^{-2} = 8^{2 \times -2} = 8^{-4} = \frac{1}{8^4} = \frac{1}{8 \times 8 \times 8 \times 8} = \frac{1}{4,096} = 0.00024$

28. $\sqrt{25} = \sqrt{5^2} = 5$

29. $\sqrt{144} = \sqrt{12^2} = 12$

30. $\sqrt{169} = \sqrt{13^2} = 13$

31. $\sqrt{100} = \sqrt{10^2} = 10$

32. $\sqrt{60} = \sqrt{4 \times 15} = \sqrt{2^2 \times 15} = 2\sqrt{15}$

33. $\sqrt{80} = \sqrt{4 \times 20} = \sqrt{4 \times 4 \times 5} = \sqrt{2^2 \times 2^2 \times 5} =$
$2 \times 2\sqrt{5} = 4\sqrt{5}$

34. $\sqrt{17} = \sqrt{17}$ (17 is a prime number)

35. $\sqrt{18} = \sqrt{9 \times 2} = \sqrt{3^2 \times 2} = 3\sqrt{2}$

36. $\sqrt{27} = \sqrt{9 \times 3} = \sqrt{3^2 \times 3} = 3\sqrt{3}$

37. $(3 \times 4 + 3^2) = (3 \times 4 + 9) = (12 + 9) = 21$

38. $(3 + 4 \times 3^2) = (3 + 4 \times 9) = (3 + 36) = 39$

39. $(4 \times 3 + 4^2) = (4 \times 3 + 16) = (12 + 16) = 28$

40. $(4 + 3 \times 4^2) = (4 + 3 \times 16) = (4 + 48) = 52$

41. $(4 \times 3 + 4^2) + (4 + 3 \times 4^2) = (12 + 16) + (4 + 3 \times 16) =$
$28 + (4 + 48) = 80$

42. $(4 \times 3^2 + 4) + (4 \times 3^2 \times 4) = (4 \times 9 + 4) + (4 \times 9 \times 4) =$
$(36 + 4) + (36 \times 4) = 40 + 144 = 184$

43. $(2 \times 2^2 + 2) \times (3 \times 3^3 + 3) = (2 \times 4 + 2) \times (3 \times 27 + 3) =$
$(8 + 2) \times (81 + 3) = 10 \times 84 = 840$

44. $6 \times 3 + (2 \div 1) \times 3 + \dfrac{3}{4} = 6 \times 3 + 2 \times 3 + \dfrac{3}{4} = 18 + 6 + \dfrac{3}{4} =$
$24\dfrac{3}{4}$

45. $\dfrac{(2 \times 2^2 + 2)}{(3 \times 3^3 + 3)} = \dfrac{(2 \times 4 + 2)}{(3 \times 27 + 3)} = \dfrac{(8 + 2)}{(81 + 3)} = \dfrac{(10)}{(84)} = \dfrac{5}{42}$

46. $\dfrac{(2 \times 2^2 + 4)}{\sqrt{2 \times 4 \times 2}} = \dfrac{(2 \times 4 + 4)}{\sqrt{16}} = \dfrac{(8 + 4)}{4} = \dfrac{12}{4} = 3$

47. $\dfrac{(4 \times 5^2)}{\sqrt{3 \times 25 + 5^2}} \times \dfrac{\sqrt{4 \times 50 + 5^2}}{(3 \times 3^2 \div 9)} = \dfrac{(4 \times 25)}{\sqrt{3 \times 25 + 25}} \times \dfrac{\sqrt{4 \times 50 + 25}}{(3 \times 9 \div 9)} = \dfrac{100}{\sqrt{75 + 25}} \times$
$\dfrac{\sqrt{200 + 25}}{(27 \div 9)} = \dfrac{100}{\sqrt{100}} \times \dfrac{\sqrt{225}}{3} = \dfrac{100}{10} \times \dfrac{15}{3} = 10 \times 5 = 50$

48. $\dfrac{\left(\frac{1}{2}\right)^3}{\sqrt{4 \times 15 + 2^4}} \div \dfrac{(8 \times 7^2)}{(3 \times 3^2 \div 9)} = \dfrac{\frac{1}{2} \times \frac{1}{2} \times \frac{1}{2}}{\sqrt{4 \times 15 + 2 \times 2 \times 2 \times 2}} \div \dfrac{(8 \times 7 \times 7)}{(3 \times 3 \times 3 \div 9)} = \dfrac{\frac{1}{8}}{\sqrt{60 + 16}} \div \dfrac{392}{(27 \div 9)} = \dfrac{\frac{1}{8}}{\sqrt{76}} \div$
$\dfrac{392}{3} = \dfrac{\frac{1}{8}}{\sqrt{4 \times 19}} \div \dfrac{392}{3} = \dfrac{\frac{1}{8}}{2\sqrt{19}} \div \dfrac{392}{3} = \dfrac{1}{8 \times 2\sqrt{19}} \div \dfrac{392}{3} = \dfrac{1}{16\sqrt{19}} \div \dfrac{392}{3} = \dfrac{1}{16\sqrt{19}} \times \dfrac{3}{392} =$
$\dfrac{3}{6{,}272\sqrt{19}} = 0.00011$

49. $\dfrac{\left(\frac{1}{2}\right)^3}{3^{-2}} + \dfrac{4^{-3}}{\left(3 \times \frac{7}{8}\right)} = \dfrac{\frac{1}{2} \times \frac{1}{2} \times \frac{1}{2}}{\frac{1}{3 \times 3}} + \dfrac{\frac{1}{4} \times \frac{1}{4} \times \frac{1}{4}}{\frac{21}{8}} = \dfrac{\frac{1}{8}}{\frac{1}{9}} + \dfrac{\frac{1}{64}}{\frac{21}{8}} = \dfrac{9}{1} \times \dfrac{1}{8} + \dfrac{8}{21} \times \dfrac{1}{64} = \dfrac{9}{8} + \dfrac{8}{1{,}344} =$
$\dfrac{168}{168} \times \dfrac{9}{8} + \dfrac{8}{1{,}344} = \dfrac{1{,}512}{1{,}344} + \dfrac{8}{1{,}344} = \dfrac{1{,}520}{1{,}344} = \dfrac{95}{84} = 1.13$

50. $\dfrac{2^2}{\left(2\times\frac{1}{3}\right)} - \dfrac{2^{-4}}{5^2} = \dfrac{4}{\frac{2}{3}} - \dfrac{\frac{1}{2}\times\frac{1}{2}\times\frac{1}{2}\times\frac{1}{2}}{25} = \dfrac{3}{2} \times 4 - \dfrac{\frac{1}{16}}{25} = \dfrac{12}{2} - \dfrac{1}{16\times25} = 6 - \dfrac{1}{400} = \dfrac{2,400}{400} -$

$\dfrac{1}{400} = \dfrac{2,399}{400} = 5\dfrac{399}{400} = 5.9975$

Answers to Chapter 2 Functions

What is the y-intercept of the below equations?

51. $y = x + 1$
Answer: $y = 1$

52. $y = 2x - 1$
Answer: $y = -1$

53. $y = 3x + 2$
Answer: $y = 2$

54. $y = -4x + 7$
Answer: $y = 7$

55. $2y = 4x + 2$
$$\frac{2y}{2} = \frac{4x}{2} + \frac{2}{2}$$
$y = 2x + 1$
Answer: $y = 1$

56. $3y = 2x + 2$
$$\frac{3y}{3} = \frac{2x}{3} + \frac{2}{3}$$
$$y = \frac{2x}{3} + \frac{2}{3}$$
Answer: $y = \frac{2}{3}$

57. $x = y - 1$
$-y = -x - 1$
$y = x + 1$
Answer: $y = 1$

58. $-x = y + 1$
$-1 - x = y$
Answer: $y = -1$

59. $4y = x + 8$
$$\frac{4y}{4} = \frac{x}{4} + \frac{8}{4}$$
$$y = \frac{x}{4} + 2$$
Answer: $y = 2$

60. $y = 5$
Answer: $y = 5$

What is the slope of the below equations?

61. $y = x + 1$

Answer: $m = 1$

62. $y = 2x - 1$

Answer: $m = 2$

63. $y = 3x + 2$

Answer: $m = 3$

64. $y = -4x + 7$

Answer: $m = -4$

65. $2y = 4x + 2$

$$\frac{2y}{2} = \frac{4x}{2} + \frac{2}{2}$$

$y = 2x + 1$

Answer: $m = 2$

66. $3y = 2x + 2$

$$\frac{3y}{3} = \frac{2x}{3} + \frac{2}{3}$$

$$y = \frac{2}{3}x + \frac{2}{3}$$

Answer: $m = \frac{2}{3}$

67. $x = y - 1$

$-y = -x - 1$

$y = x + 1$

Answer: $m = 1$

68. $-x = y + 1$

$-y = x + 1$

$$\frac{-y}{-1} = \frac{x}{-1} + \frac{1}{-1}$$

$y = -x - 1$

Answer: $m = -1$

69. $4y = x + 8$

$$\frac{4y}{4} = \frac{x}{4} + \frac{8}{4}$$

$$y = \frac{x}{4} + 2$$

Answer: $m = \frac{1}{4}$

70. $y = 5$

Answer: $m = 0$

Write the equation of the below graphs in y-intercept form:

71. *Answer:* $y = x + 3$

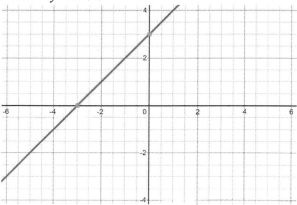

72. *Answer:* $y = 2x + 3$

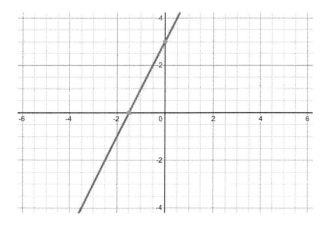

73. *Answer:* $y = 4x + 5$

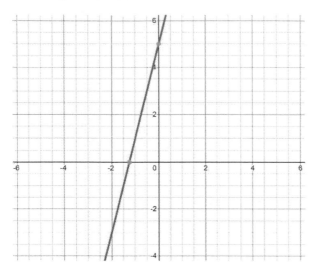

74. *Answer:* $y = -4x + 5$

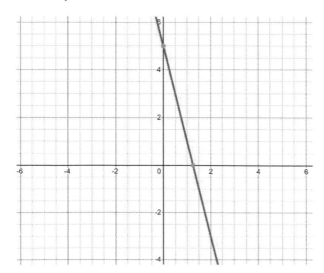

75. *Answer:* $y = \dfrac{-2}{3}x + 3$

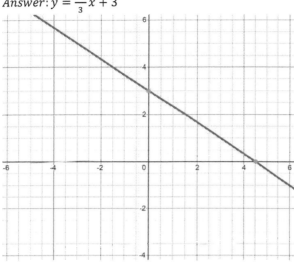

76. *Answer:* $y = \dfrac{x}{4} - 3$

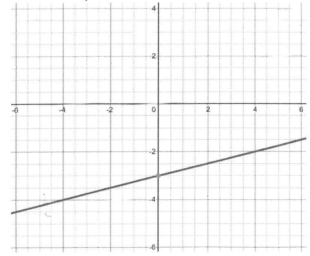

77. *Answer:* $y = \frac{x}{2} - 2$

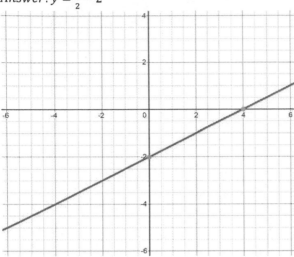

78. *Answer:* $y = \frac{1}{2}$

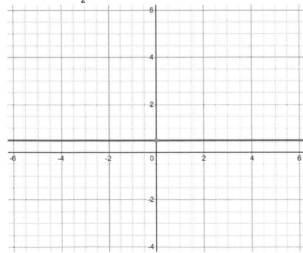

79. *Answer:* $y = x$

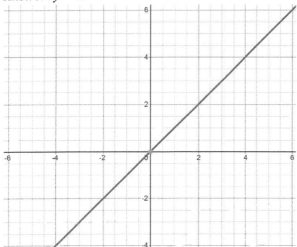

80. *Answer:* $y = -2x$

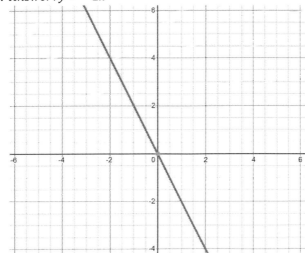

Indicate if the below graphs represent a function:

81. *Answer:* yes
82. *Answer:* no

83. *Answer*: *no*
84. *Answer*: *yes*
85. *Answer*: *yes*

Answers to Chapter 3 Solving Equations

Solve for x:

86. $4(4x - 3) = 2(7x - 1)$

$16x - 12 = 14x - 2$

$16x - 14x = -2 + 12$

$2x = 10$

$$\frac{2x}{2} = \frac{10}{2}$$

$x = 5$

87. $2(5x - 6) = 8(x - 4)$

$10x - 12 = 8x - 32$

$10x - 8x = -32 + 12$

$2x = -20$

$$\frac{2x}{2} = \frac{-20}{2}$$

$x = -10$

88. $2(4x - 4) = 3(8x - 12)$

$8x - 8 = 24x - 36$

$8x - 24x = -36 + 8$

$-16x = -28$

$$\frac{-16x}{-16} = \frac{-28}{-16}$$

$$x = \frac{24}{16} = \frac{3}{2}$$

89. $4x + 3 = 3x - 4$

$4x - 3x = -4 - 3$

$x = -7$

90. $7x + 14 = 21x + 28$

$7x - 21x = 28 - 14$

$-14x = 14$

$x = -1$

91. $x + 2 = 3x - 4$

$x - 3x = -4 - 2$

$-2x = -6$

$$\frac{-2x}{-2} = \frac{-6}{-2}$$

$x = 3$

92. $12 + 2x = 8$

$2x = 8 - 12$

$2x = -4$

$$\frac{2x}{2} = \frac{-4}{2}$$

$x = -2$

93. $3(x + 1) = 6x - 4$

$3x + 3 = 6x - 4$

$3x - 6x = -4 - 3$

$-3x = -7$

$$\frac{-3x}{-3} = \frac{-7}{-3}$$

$$x = \frac{7}{3}$$

94. $2 - x = x - 2$

$-x - x = -2 - 2$

$-2x = -4$

$$\frac{-2x}{-2} = \frac{-4}{-2}$$

$x = 2$

95. $3x + 9 = 7 + x$

$3x - x = 7 - 9$

$2x = -2$

$$\frac{2x}{2} = \frac{-2}{2}$$

$x = -1$

96. $9x - 9 = 21 + 7x$

$9x - 7x = 21 + 9$

$2x = 30$

$$\frac{2x}{2} = \frac{30}{2}$$

$x = 15$

97. $(3 - 4x)5 = 20 - 15x$

$15 - 20x = 20 - 15x$

$-20x + 15x = 20 - 15$

$-5x = 5$

$$\frac{-5x}{-5} = \frac{5}{-5}$$

$x = -1$

98. $9x + 8 = 4x - 3$

$$9x - 4x = -3 - 8$$
$$5x = -11$$
$$\frac{5x}{5} = \frac{-11}{5}$$
$$x = -\frac{11}{5}$$

99. $5x - 5 = 4x + 17$
$$5x - 4x = 17 + 5$$
$$x = 22$$

100. $12x + 15 = 15x - 18$
$$12x - 15x = -18 - 15$$
$$-3x = -33$$
$$\frac{-3x}{-3} = \frac{-33}{-3}$$
$$x = 11$$

101. $13(2x - 8) = 7(8x + 7)$
$$26x - 104 = 56x + 49$$
$$-30x = 153$$
$$\frac{-30x}{-30} = \frac{153}{-30}$$
$$x = -\frac{153}{30} = -\frac{51}{10}$$

102. $15x + 20 = 12x + 25$
$$15x - 12x = 25 - 20$$
$$3x = 5$$
$$\frac{3x}{3} = \frac{5}{3}$$
$$x = \frac{5}{3}$$

103. $7x + 14 = 21x + 28$
$$7x - 21x = 28 - 14$$
$$-14x = 14$$
$$\frac{-14x}{-14} = \frac{14}{-14}$$
$$x = -1$$

104. $13 + 17x = 21$
$$17x = 21 - 13$$
$$17x = 8$$
$$\frac{17x}{17} = \frac{8}{17}$$

$$x = \frac{8}{17}$$

105. $12(4x - 5) = 11(8x - 12)$
$$48x - 60 = 88x - 132$$
$$-40x = -72$$
$$\frac{-40x}{-40} = \frac{-72}{-40}$$
$$x = \frac{72}{40} = \frac{9}{5}$$

106. $14(9x - 7) = 13x + 18$
$$126x - 98 = 13x + 18$$
$$126x - 13x = 18 + 98$$
$$113x = 116$$
$$x = \frac{113}{116}$$

107. $22(8 - 3x) = 12x + 46$
$$176 - 66x = 12x + 46$$
$$-66x - 12x = 46 - 176$$
$$-78x = -130$$
$$\frac{-78x}{-78} = \frac{-130}{-78}$$
$$x = \frac{130}{78} = \frac{65}{39}$$

108. $12(7x - 7) = 20x$
$$84x - 84 = 20x$$
$$84x - 20x = 84$$
$$64x = 84$$
$$\frac{64x}{64} = \frac{84}{64}$$
$$x = \frac{21}{16}$$

109. $2(4x - 2) = 32x - 2$
$$8x - 4 = 32x - 2$$
$$8x - 32x = -2 + 4$$
$$-24x = 2$$
$$\frac{-24x}{-24} = \frac{2}{-24}$$
$$x = -\frac{1}{12}$$

110. $8(3x - 9) = 4x + 18$
$$24x - 72 = 4x + 18$$

$$24x - 4x = 18 + 72$$
$$20x = 90$$
$$\frac{20x}{20} = \frac{90}{20}$$
$$x = \frac{9}{2}$$

111. $16 + 6x = 41 + x$
$$16 - 41 = x - 6x$$
$$-25 = -5x$$
$$\frac{-25}{-5} = \frac{-5x}{-5}$$
$$x = 5$$

112. $x + 27 = 3x + 32$
$$x - 3x = 32 - 27$$
$$-2x = 5$$
$$\frac{-2x}{-2} = \frac{5}{-2}$$
$$x = -\frac{5}{2}$$

113. $18x + 21 = 26x - 35$
$$18x - 26x = -35 - 21$$
$$-8x = -56$$
$$\frac{-8x}{-8} = \frac{-56}{-8}$$
$$x = 7$$

114. $19x + 81 = 42x - 34$
$$19x - 42x = -34 - 81$$
$$-23x = -115$$
$$\frac{-23x}{-23} = \frac{-115}{-23}$$
$$x = 5$$

115. $42x - 34 = 32 + 15x$
$$42x - 15x = 32 + 34$$
$$27x = 66$$
$$\frac{27x}{27} = \frac{66}{27}$$
$$x = \frac{22}{9}$$

116. $27 - 7x = 3x - 3$
$$27 + 3 = 3x + 7x$$
$$30 = 10x$$

$$\frac{30}{10} = \frac{10x}{10}$$
$$3 = x$$

117. $12 + 12x = 36$
$$12x = 36 - 12$$
$$12x = 24$$
$$\frac{12x}{12} = \frac{24}{12}$$
$$x = 2$$

118. $x + 1 = 2x + 28$
$$x - 2x = 28 - 1$$
$$-x = 27$$
$$x = -27$$

119. $3(6x - 7) = 2(8x - 14)$
$$18x - 21 = 16x - 28$$
$$18x - 16x = -28 + 21$$
$$2x = -7$$
$$\frac{2x}{2} = \frac{-7}{2}$$
$$x = -\frac{7}{2}$$

120. $5(2x - 7) = 8(7x - 21)$
$$10x - 35 = 56x - 168$$
$$10x - 56x = -168 + 35$$
$$-46x = -133$$
$$\frac{-46x}{-46} = \frac{-133}{-46}$$
$$x = \frac{133}{46}$$

121. $12(10x + 10) = 6(5x - 13)$
$$120x + 120 = 30x - 78$$
$$120x - 30x = -78 - 120$$
$$90x = -198$$
$$\frac{90x}{90} = \frac{-198}{90}$$
$$x = -\frac{198}{90} = -\frac{11}{5}$$

122. $x + 1 = -x - 2$
$$x + x = -2 - 1$$
$$2x = -3$$

$$\frac{2x}{2} = -\frac{3}{2}$$

$$x = -\frac{3}{2}$$

123. $3x + 12 = -4x - 9$

$$3x + 4x = -9 - 12$$

$$7x = -21$$

$$\frac{7x}{7} = \frac{-21}{7}$$

$$x = -3$$

124. $7(2x + 6) = 16x - 12$

$$14x + 42 = 16x - 12$$

$$-16x + 14x = -12 - 42$$

$$-2x = -54$$

$$\frac{-2x}{-2} = \frac{-54}{-2}$$

$$x = 27$$

125. $12x - 20 = 15 - 8x$

$$12x + 8x = 15 + 20$$

$$20x = 35$$

$$\frac{20x}{20} = \frac{35}{20}$$

$$x = \frac{35}{20} = \frac{7}{4}$$

126. $(1 - x)12 = 5 - 12x$

$$12 - 12x = 5 - 12x$$

$$-12x + 12x = 5 - 12$$

$$0x = -7$$

$$x = \infty$$

127. $51x - 52 = 34x + 16$

$$51x - 34x = 16 + 52$$

$$17x = 68$$

$$\frac{17x}{17} = \frac{68}{17}$$

$$x = 4$$

128. $7(2x - 4) = 9(6x + 8)$

$$14x - 28 = 54x + 72$$

$$14x - 54x = 72 + 28$$

$$-40x = 100$$

$$\frac{-40x}{-40} = \frac{100}{-40}$$

$$x = -\frac{100}{40} = -\frac{5}{2}$$

129. $4x + 24 = 31x + 54$

$$4x - 31x = 54 - 24$$

$$-27x = 30$$

$$\frac{-27x}{-27} = \frac{30}{-27}$$

$$x = -\frac{30}{27} = -\frac{10}{9}$$

130. $6(5x - 8) = 12(4x - 8)$

$$30x - 48 = 48x - 96$$

$$30x - 48x = -96 + 48$$

$$-18x = -48$$

$$\frac{18x}{-18} = \frac{-48}{-18}$$

$$x = \frac{8}{3}$$

131. $4(22 - 12x) = 12x + 44$

$$88 - 48x = 12x + 44$$

$$88 - 44 = 12x + 48x$$

$$44 = 60x$$

$$\frac{44}{60} = \frac{60x}{60}$$

$$\frac{11}{15} = \frac{44}{60} = x$$

132. $8(7x - 9) = 12(10x - 7)$

$$56x - 72 = 120x - 84$$

$$56x - 120x = -84 + 72$$

$$-64x = -12$$

$$\frac{-64x}{-64} = \frac{-12}{-64}$$

$$x = \frac{12}{64} = \frac{3}{16}$$

133. $4(4x - 7) = 9(9x + 8)$

$$16x - 28 = 81x + 72$$

$$16x - 81x = 72 + 28$$

$$-65x = 100$$

$$\frac{-65x}{-65} = \frac{100}{-65}$$

$$x = -\frac{100}{65} = -\frac{20}{13}$$

134. $(7-x)5 = 9 - 11x$

$$35 - 5x = 9 - 11x$$
$$35 - 9 = -11x + 5x$$
$$26 = -6x$$
$$\frac{26}{-6} = \frac{-6x}{-6}$$
$$-\frac{13}{3} = -\frac{26}{6} = x$$

135. $5(2x+5) = 3x - 10$

$$10x + 25 = 3x - 10$$
$$10x - 3x = -10 - 25$$
$$7x = -35$$
$$\frac{7x}{7} = \frac{-35}{7}$$
$$x = -\frac{35}{7} = -5$$

136. $9x + 12 = 8x - 4$

$$9x - 8x = -4 - 12$$
$$x = -16$$

137. $6(3x+8) = 2(6x+19)$

$$18x + 48 = 12x + 38$$
$$18x - 12x = 38 - 48$$
$$6x = -10$$
$$\frac{6x}{6} = -\frac{10}{6}$$
$$x = -\frac{10}{6} = -\frac{5}{3}$$

138. $3x + 8 = 9x + 27$

$$3x - 9x = 27 - 8$$
$$-6x = 19$$
$$\frac{-6x}{-6} = \frac{19}{-6}$$
$$x = -\frac{19}{6}$$

139. $24 + 5x = 8x + 3$

$$24 - 3 = 8x - 5x$$
$$21 = 3x$$
$$\frac{21}{3} = \frac{3x}{3}$$
$$7 = x$$

140. $12x - 30 = 46x + 38$

$$12x - 46x = 38 + 30$$
$$-34x = 68$$
$$\frac{-34x}{-34} = \frac{68}{-34}$$
$$x = -2$$

141. $2x + 18 = 6x + 24$

$$2x - 6x = 24 - 18$$
$$-4x = 6$$
$$\frac{-4x}{-4} = \frac{6}{-4}$$
$$x = -\frac{3}{2}$$

142. $12(4x-4) = 6x + 18$

$$48x - 48 = 6x + 18$$
$$48x - 6x = 18 + 48$$
$$42x = 66$$
$$\frac{42x}{42} = \frac{66}{42}$$
$$x = \frac{33}{21} = \frac{11}{7}$$

143. $3(9x-6) = 3x - 12$

$$27x - 18 = 3x - 12$$
$$27x - 3x = -12 + 18$$
$$24x = 6$$
$$\frac{24x}{24} = \frac{6}{24}$$
$$x = \frac{6}{24} = \frac{1}{4}$$

144. $14 + 16x = 42 + 2x$

$$14 - 42 = 2x - 16x$$
$$-28 = -14x$$
$$\frac{-28}{-14} = \frac{-14x}{-14}$$
$$2 = x$$

145. $28x + 41 = 56x - 65$

$$28x - 56x = -65 - 41$$
$$-28x = -106$$
$$\frac{-28x}{-28} = \frac{-106}{-28}$$
$$x = \frac{106}{28} = 3\frac{11}{14}$$

146. $32x - 94 = 62 + 35x$

$$32x - 35x = 94 + 62$$
$$-3x = 156$$
$$\frac{-3x}{-3} = \frac{156}{-3}$$
$$x = -52$$

147. $4 - 2x = 6x + 72$

$$-2x - 6x = 72 - 4$$
$$-8x = 68$$
$$-\frac{8x}{8} = -\frac{68}{8}$$
$$x = -\frac{68}{8} = -\frac{34}{4} = -\frac{17}{2}$$

148. $2(6x + 9) = 2(3x + 16)$

$$12x + 18 = 6x + 32$$
$$12x - 6x = 32 - 18$$
$$6x = 14$$
$$\frac{6x}{6} = \frac{14}{6}$$
$$x = \frac{7}{3}$$

149. $2(20x + 30) = 4(4x - 12)$

$$40x + 60 = 16x - 48$$
$$40x - 16x = -48 - 60$$
$$24x = -108$$
$$\frac{24x}{24} = -\frac{108}{24}$$
$$x = -\frac{54}{12} = -\frac{27}{6} = -\frac{9}{2}$$

150. $-24x + 120 = 15x + 90$

$$-24x - 15x = 90 - 120$$
$$-39x = -30$$
$$\frac{-39x}{-39} = \frac{-30}{-39}$$
$$x = \frac{30}{39} = \frac{10}{13}$$

151. $|4 + x| = 12$

$$(4 + x) = \pm 12$$
$$(4 + x) = 12$$
$$x = 12 - 4$$
$$x = 8$$

$$(4 + x) = -12$$
$$x = -12 - 4$$
$$x = -16$$
$$x = 8, -16$$

152. $|2 + x| = 6$

$$(2 + x) = \pm 6$$
$$(2 + x) = 6$$
$$x = 6 - 2$$
$$x = 4$$
$$(2 + x) = -6$$
$$x = -6 - 2$$
$$x = -8$$
$$x = 4, -8$$

153. $|3 + x| = 9$

$$(3 + x) = \pm 9$$
$$x = -9 - 3$$
$$x = -12$$
$$x = 9 - 3$$
$$x = 6$$
$$x = -12, 6$$

154. $2|2 + x| = 8$

$$\frac{|2 + x|}{2} = \frac{8}{2}$$
$$|2 + x| = 4$$
$$(2 + x) = \pm 4$$
$$(2 + x) = 4$$
$$x = 4 - 2$$
$$x = 2$$
$$(2 + x) = -4$$
$$x = -4 - 2$$
$$x = -6$$
$$x = 2, -6$$

155. $4|3 + x| = 16$

$$\frac{|3 + x|}{4} = \frac{16}{4}$$
$$|3 + x| = 4$$
$$(3 + x) = \pm 4$$
$$3 + x = 4$$
$$x = 4 - 3$$

$$x = 1$$
$$3 + x = -4$$
$$x = -4 - 3$$
$$x = -7$$
$$x = 1, -7$$

156. $|x + 1| = 7$
$$(x + 1) = \pm 7$$
$$x + 1 = 7$$
$$x = 7 - 1$$
$$x = 6$$
$$x + 1 = -7$$
$$x = -7 - 1$$
$$x = -8$$
$$x = 6, -8$$

157. $6|x + 12| = 2$
$$\frac{|x + 12|}{6} = \frac{2}{6}$$
$$|x + 12| = \frac{1}{3}$$
$$(x + 12) = \pm\frac{1}{3}$$
$$x = -12 + \frac{1}{3} = -11\frac{2}{3}$$
$$x = -12 - \frac{1}{3} = -12\frac{1}{3}$$
$$x = -12\frac{1}{3}, -11\frac{2}{3}$$

158. $8|3 + x| = 24$
$$\frac{|3 + x|}{8} = \frac{24}{8}$$
$$|3 + x| = 3$$
$$(3 + x) = \pm 3$$
$$3 + x = 3$$
$$x = 3 - 3$$
$$x = 0$$

$$3 + x = -3$$
$$x = -3 - 3$$
$$x = -6$$
$$x = 0, -6$$

159. $2|x + 2| = 12$
$$\frac{|x + 2|}{2} = \frac{12}{2}$$
$$|x + 2| = 6$$
$$(x + 2) = \pm 6$$
$$x = 6 - 2$$
$$x = 4$$
$$x = -6 - 2$$
$$x = -8$$
$$x = 4, -8$$

160. $8|x + 2| = 2$
$$\frac{|x + 2|}{8} = \frac{2}{8}$$
$$|x + 2| = \frac{1}{4}$$
$$(x + 2) = \pm\frac{1}{4}$$
$$(x + 2) = \frac{1}{4}$$
$$x = \frac{1}{4} - 2$$
$$x = -1\frac{3}{4}$$
$$(x + 2) = -\frac{1}{4}$$
$$x = -\frac{1}{4} - 2$$
$$x = -2\frac{1}{4}$$
$$x = -1\frac{3}{4}, -2\frac{1}{4}$$

Combine like terms:

161. $2y + 3x + 4y + x$
$$2y + 4y + 3x + x$$

$$6y + 4x$$

162. $y + x - 2y + 3x$

$$y - 2y + x + 3x$$
$$-y + 4x$$

163. $7y + 7x - 3y + 3x$

$$7y - 3y + 7x + 3x$$
$$4y + 10x$$

164. $2y + 3x + 4y + 6x$

$$2y + 4y + 3x + 6x$$
$$6y + 9x$$

165. $2xy + 2y + 2xy$

$$2xy + 2xy + 2y$$
$$4xy + 2y$$

166. $3xy + 7x + 4xy - y + x$

$$3xy + 4xy + 7x + x - y$$
$$7xy + 8x - y$$

167. $2xy + 3x - 3y + 4 + 5xy - 12$

$$2xy + 5xy + 3x - 3y + 4 - 12$$
$$7xy + 3x - 3y - 8$$

168. $2y + 3x - 10y + 4x - 5y - 3y + 3xy - 2x$

$$2y - 10y - 5y - 3y + 3x + 4x - 2x + 3xy$$
$$-16y + 5x + 3xy$$

169. $x + 6x - 6xy - 5y + 5y - 9x + 8z - 9z$

$$x + 6x - 9x - 5y + 5y - 6xy + 8z - 9z$$
$$-2x - 6xy - z$$

170. $6y + 5y - 8z - xy - 5y - 6z + xy - 4y$

$$6y + 5y - 5y - 4y - 8z - 6z - xy + xy$$
$$2y - 14z$$

171. $8xy + 7xy + 6xy - 4 - 5y + 4x - 3x$

$$21xy - 4 - 5y + x$$

172. $3z + 8xy - 7x + 9z + 5y - 4xy - 6$

$$3z + 9z + 8xy - 4xy - 7x + 5y - 6$$
$$12z + 4xy - 7x + 5y - 6$$

173. $9xz + 3z + 6y - 4 + 5y - 2x + 9z - 9$

$$9xy + 3z + 9z + 6y + 5y - 4 - 9 - 2x$$
$$9xy + 12z + 11y - 13 - 2x$$

174. $6x - 9xy - 2 + 5y - y + 7xy - 3z$

$$6x - 9xy + 7xy - 2 + 5y - y - 3z$$
$$6x - 2xy - 2 + 4y - 3z$$

175. $2 + 3y + 4z - 6 - 5y - 6x + 4z - 2y$
$$2 - 6 + 3y - 5y - 2y + 4z + 4z - 6x$$
$$-4 - 4y + 8z - 6x$$

176. $6x + 2z + 2y - 7x + 5xy + 8z + 2xy$
$$6x - 7x + 2z + 8z + 2y + 5xy + 2xy$$
$$-x + 10z + 2y + 7xy$$

177. $z - 4x + 2xy - x - z - 3xy + 2z$
$$z - z + 2z - 4x - x + 2xy - 3xy$$
$$2z - 5x - xy$$

178. $z + 3xy - y + 5y + 2y - 7xy$
$$z + 3xy - 7xy - y + 5y + 2y$$
$$z - 4xy + 6y$$

179. $xz + 3z + 6y + 7z + -2xz + y - 2x$
$$xz - 2xz + 3z + 7z + 6y + y - 2x$$
$$-xz + 10z + 7y - 2x$$

180. $xy + y + 7z - 6 + 8xy - 6z + 2y - 4xy$
$$xy + 8xy - 4xy + y + 2y + 7z - 6z - 6$$
$$5xy + 3y + z - 6$$

Answers to Chapter 4 Inequalities

Solve for x:

181. $x + 3 > 2$

$x > 2 - 3$

$x > -1$

182. $x + 7 < 4$

$x < 4 - 7$

$x < -3$

183. $x - 4 < -7$

$x < -7 + 4$

$x < -3$

184. $x + 2 > 1$

$x > 1 - 2$

$x > -1$

185. $x - 2 < 9$

$x < 9 + 2$

$x < 11$

186. $-1 + x < 2$

$x < 2 + 1$

$x < 3$

187. $x + 6 > 8$

$x > 8 - 6$

$x > 2$

188. $-5 + x < -1$

$x < -1 + 5$

$x < 4$

189. $x + 3 \geq -4$

$x \geq -4 - 3$

$x \geq -7$

190. $7 + x \leq 6$

$x \leq 6 - 7$

$x \leq -1$

191. $6 + x < 9$

$x < 9 - 6$

$x < 3$

192. $-7 + x \geq -4$

$x \geq -4 + 7$

$x \geq 3$

193. $x + 7 < 4$

$x < 4 - 7$

$x < -3$

194. $-5 + x \geq -5$

$x \geq -5 + 5$

$x \geq 0$

195. $x - 1 > 1$

$x > 1 + 1$

$x > 2$

196. $x + 1 \geq 1$

$x \geq 1 - 1$

$x \geq 0$

197. $x + 8 < 3$

$x < 3 - 8$

$x < -5$

198. $x - 5 \leq 10$

$x \leq 10 + 5$

$x \leq 15$

199. $-2 + x < 7$

$x < 7 + 2$

$x < 9$

200. $x + 6 > 1$

$x > 1 - 6$

$x > -5$

201. $2x > 2$

$\dfrac{2x}{2} > \dfrac{2}{2}$

$x > 1$

202. $3x < -6$

$\dfrac{3x}{3} < \dfrac{-6}{3}$

$x < -2$

203. $7x > 14$

$$\frac{7x}{7} > \frac{14}{7}$$
$$x > 2$$

204. $12x > 6$

$$\frac{12x}{12} > \frac{6}{12}$$
$$x > \frac{1}{2}$$

205. $3x \geq 12$

$$\frac{3x}{3} \geq \frac{12}{3}$$
$$x \geq 4$$

206. $2x \leq -4$

$$\frac{2x}{2} \leq \frac{-4}{2}$$
$$x \leq -2$$

207. $3x \geq 2$

$$\frac{3x}{3} \geq \frac{2}{3}$$
$$x \geq \frac{2}{3}$$

208. $3x > 9$

$$\frac{3x}{3} > \frac{9}{3}$$
$$x > 3$$

209. $6x \leq 15$

$$\frac{6x}{6} \leq \frac{15}{6}$$
$$x \leq \frac{5}{2}$$

210. $-2x > 6$

$$\frac{-2x}{-2} > \frac{6}{-2}$$
$$x < -3$$

211. $-x \leq 2$

$$\frac{-x}{-1} \leq \frac{2}{-1}$$
$$x \geq -2$$

212. $-x < -2$

$$\frac{-x}{-1} < \frac{-2}{-1}$$
$$x > 2$$

213. $-6x \geq 15$

$$\frac{-6x}{-6} \geq \frac{15}{-6}$$
$$x \leq -\frac{5}{2}$$

214. $2x > -5$

$$\frac{2x}{2} > \frac{-5}{2}$$
$$x > -\frac{5}{2}$$

215. $4x \leq 12$

$$\frac{4x}{4} \leq \frac{12}{4}$$
$$x \leq 3$$

216. $3x > 9$

$$\frac{3x}{3} > \frac{9}{3}$$
$$x > 3$$

217. $2x \geq 8$

$$\frac{2x}{2} \geq \frac{8}{2}$$
$$x \geq 4$$

218. $-x < 2$

$$\frac{-x}{-1} < \frac{2}{-1}$$
$$x > -2$$

219. $-3x \leq -6$

$$\frac{-3x}{-3} \leq \frac{-6}{-3}$$
$$x \geq 2$$

220. $5x \geq -15$

$$\frac{5x}{5} \geq \frac{-15}{5}$$
$$x \geq -3$$

221. $2(x + 1) \geq 4$

$$2x + 2 \geq 4$$
$$2x \geq 4 - 2$$
$$2x \geq 2$$
$$\frac{2x}{2} \geq \frac{2}{2}$$
$$x \geq 1$$

222. $5(x - 2) \leq 2$

$$5x - 10 \leq 2$$
$$5x \leq 2 + 10$$
$$5x \leq 12$$
$$\frac{5x}{5} \leq \frac{12}{5}$$
$$x \leq \frac{12}{5}$$

223. $3(x + 2) \leq 12$

$$3x + 6 \leq 12$$
$$3x \leq 12 - 6$$
$$3x \leq 6$$
$$\frac{3x}{3} \leq \frac{6}{3}$$
$$x \leq 2$$

224. $7(2x - 1) > -14$

$$14x - 7 > -14$$
$$14x > -14 + 7$$
$$14x > -7$$
$$\frac{14x}{14} > \frac{-7}{14}$$
$$x > -\frac{1}{2}$$

225. $-2(x - 3) \geq -12$

$$-2x + 6 \geq -12$$
$$-2x \geq -12 - 6$$
$$-2x \geq -18$$
$$\frac{-2x}{-2} \geq \frac{-18}{-2}$$
$$x \leq 9$$

226. $-(x + 2) > -4$

$$\frac{-(x + 2)}{-1} > \frac{-4}{-1}$$
$$x + 2 < 4$$
$$x < 4 - 2$$
$$x < 2$$

227. $4(x + 1) < 16$

$$\frac{4(x + 1)}{4} < \frac{16}{4}$$
$$x + 1 < 4$$
$$x < 4 - 1$$

$x < 3$

228. $-3(x + 1) < 4$

$$-3x - 3 < 4$$
$$-3x < 4 + 3$$
$$-3x < 7$$
$$\frac{-3x}{-3} < \frac{7}{-3}$$
$$x > -\frac{7}{3}$$

229. $8(x - 1) \geq -4$

$$8x - 8 \geq -4$$
$$8x \geq -4 + 8$$
$$8x \geq 4$$
$$\frac{8x}{8} \geq \frac{4}{8}$$
$$x \geq \frac{1}{2}$$

230. $5(2x - 3) \leq 25$

$$10x - 15 \leq 25$$
$$10x \leq 25 + 15$$
$$10x \leq 40$$
$$\frac{10x}{10} \leq \frac{40}{10}$$
$$x \leq 4$$

231. $5(x - 1) \leq 5(2x - 3)$

$$5x - 5 \leq 10x - 15$$
$$5x - 10x \leq -15 + 5$$
$$-5x \leq -10$$
$$\frac{-5x}{-5} \leq \frac{-10}{-5}$$
$$x \geq 2$$

232. $4(x + 2) < 3(5x + 7)$

$$4x + 8 < 15x + 21$$
$$4x - 15x < 21 - 8$$
$$-11x < 13$$
$$\frac{-11x}{-11} < \frac{13}{-11}$$
$$x > -\frac{13}{11}$$

233. $5(3x + 4) \leq 5(2x + 9)$

$$15x + 20 \leq 10x + 45$$

$$15x - 10x \le 45 - 20$$
$$5x \le 25$$
$$\frac{5x}{5} \le \frac{25}{5}$$
$$x \le 5$$

234. $6(10x - 5) \ge 5(6x - 2)$
$$60x - 30 \ge 30x - 10$$
$$60x - 30x \ge -10 + 30$$
$$30x \ge 20$$
$$\frac{30x}{30} \ge \frac{20}{30}$$
$$x \ge \frac{2}{3}$$

235. $-7(7x + 2) \le 5(4x - 6)$
$$-49x - 14 \le 20x - 30$$
$$-49x - 20x \le -30 + 14$$
$$-69x \le -16$$
$$\frac{-69x}{-69} \le \frac{-16}{-69}$$
$$x \ge \frac{16}{69}$$

236. $2(15x - 3) < 5(x - 4)$
$$30x - 6 < 5x - 20$$
$$30x - 5x < -20 + 6$$
$$25x < -14$$
$$\frac{25x}{25} < -\frac{14}{25}$$

237. $4(2x - 2) > (6x + 5)$
$$8x - 8 > 6x + 5$$
$$8x - 6x > 5 + 8$$
$$2x > 13$$
$$\frac{2x}{2} > \frac{13}{2}$$
$$x > \frac{13}{2}$$

238. $(6x + 3) > 5(7x - 2)$
$$6x + 3 > 35x - 10$$
$$6x - 35x > -10 - 3$$
$$-29x > -13$$
$$\frac{-29x}{-29} > \frac{-13}{-29}$$

$$x > \frac{13}{29}$$

239. $-3(2x + 1) \le 8(x + 4)$
$$-6x - 3 \le 8x + 32$$
$$-6x - 8x \le 32 + 3$$
$$-14x \le 35$$
$$\frac{-14x}{-14} \le \frac{35}{-14}$$
$$x \ge -\frac{35}{14}$$
$$x \ge -\frac{5}{2}$$

240. $8(8x + 4) \ge 6(3x + 5)$
$$64x + 32 \ge 18x + 30$$
$$64x - 18x \ge 30 - 32$$
$$46x \ge -2$$
$$\frac{46x}{46} \ge \frac{-2}{46}$$
$$x \ge -\frac{1}{23}$$

241. $3(2x - 6) \le 2(8x - 3)$
$$6x - 18 \le 16x - 6$$
$$6x - 16x \le -6 + 18$$
$$-10x \le 12$$
$$\frac{-10x}{-10} \le \frac{12}{-10}$$
$$x \ge -\frac{6}{5}$$

242. $-5(2x + 3) \ge -10(3x - 2)$
$$-10x - 15 \ge -30x + 20$$
$$-10x + 30x \ge 20 + 15$$
$$20x \ge 35$$
$$\frac{20x}{20} \ge \frac{35}{20}$$
$$x \ge \frac{7}{4}$$

243. $9(9x - 5) \ge 6(6x + 1)$
$$81x - 45 \ge 36x + 6$$
$$81x - 36x \ge 6 + 45$$
$$45x \ge 51$$

$$\frac{45x}{45} \geq \frac{51}{45}$$

$$x \geq \frac{51}{45} \geq \frac{17}{15}$$

244. $6(2x - 10) < 7(7x + 6)$

$$12x - 60 < 49x + 42$$

$$12x - 49x < 42 + 60$$

$$-37x < 102$$

$$\frac{-37x}{-37} < \frac{102}{-37}$$

$$x > -\frac{102}{37}$$

245. $7(2x + 7) \leq 9(8x + 10)$

$$14x + 49 \leq 72x + 90$$

$$14x - 72x \leq 90 - 49$$

$$-58x \leq 41$$

$$\frac{58x}{-58} \leq \frac{41}{-58}$$

$$x \geq -\frac{41}{58}$$

246. $4(6x - 6) > -2(7x - 8)$

$$24x - 24 > -14x + 16$$

$$24x + 14x > 16 + 24$$

$$38x > 40$$

$$\frac{38x}{38} > \frac{40}{38}$$

$$x > \frac{20}{19}$$

247. $-9(2x + 9) < 5(2x + 7)$

$$-18x - 81 < 10x + 35$$

$$-18x - 10x < 35 + 81$$

$$-28x < 116$$

$$\frac{-28x}{-28} < \frac{116}{-28}$$

$$x > -\frac{29}{7}$$

248. $2(2x - 7) \leq (x - 9)$

$$4x - 14 \leq x - 9$$

$$4x - x \leq -9 + 14$$

$$3x \leq 5$$

$$\frac{3x}{3} \leq \frac{5}{3}$$

$$x \leq \frac{5}{3}$$

249. $3(4x - 3) > 10(x + 2)$

$$12x - 9 > 10x + 20$$

$$12x - 10x > 20 + 9$$

$$2x > 29$$

$$\frac{2x}{2} > \frac{29}{2}$$

$$x > \frac{29}{2}$$

250. $(2x + 8) \geq -5(9x + 8)$

$$2x + 8 \geq -45x - 40$$

$$2x + 45x \geq -40 - 8$$

$$47x \geq -48$$

$$\frac{47x}{47} \geq -\frac{48}{47}$$

$$x \geq -\frac{48}{47}$$

251. $-2 < x + 1 < 4$

$$-2 < x + 1 \quad AND \quad x + 1 < 4$$
$$-2 - 1 < x \quad AND \quad x < 4 - 1$$
$$-3 < x \quad AND \quad x < 3$$

252. $-1 > x + 3 > 3$

$$-1 > x + 3 \quad OR \quad x + 3 > 3$$
$$-3 - 1 > x \quad OR \quad x > 3 - 3$$
$$-4 > x \quad OR \quad x > 0$$

253. $5 > x + 5 > 12$

$$5 > x + 5 \quad OR \quad x + 5 > 12$$
$$5 - 5 > x \quad OR \quad x > 12 - 5$$
$$0 > x \quad OR \quad x > 7$$

254. $-2 < x + 1 < -4$

$$-2 < x + 1 \quad OR \quad x + 1 < -4$$
$$-2 - 1 < x \quad OR \quad x < -4 - 1$$
$$-3 < x \quad OR \quad x < -5$$

255. $1 < x + 7 > 4$

$$1 < x + 7 \quad AND \quad x + 7 > 4$$
$$1 - 7 < x \quad AND \quad x > 4 - 7$$
$$-6 < x \quad AND \quad x > -3$$

256. $5 > x + 7 < 7$

$$5 > x + 7 \quad AND \quad x + 7 < 7$$
$$-7 + 5 > x \quad AND \quad x < 7 - 7$$
$$-2 > x \quad AND \quad x < 0$$

257. $4 < x + 1 < 5$

$$4 < x + 1 \quad AND \quad x + 1 < 5$$
$$-1 + 4 < x \quad AND \quad x < 5 - 1$$
$$3 < x \quad AND \quad x < 4$$

258. $-18 > x + 9 > 27$

$$-18 > x + 9 \quad OR \quad x + 9 > 27$$
$$-18 - 9 > x \quad OR \quad x > 27 - 9$$
$$-27 > x \quad OR \quad x > 18$$

259. $7 < x + 7 < 8$

$$7 < x + 7 \quad AND \quad x + 7 < 8$$
$$7 - 7 < x \quad AND \quad x < 8 - 7$$
$$0 < x \quad AND \quad x < 1$$

260. $5 > x + 3 > 2$

$$5 > x + 3 \quad AND \quad x + 3 > 2$$

$$5 - 3 > x \quad AND \quad x > 2 - 3$$
$$2 > x \quad AND \quad x > -1$$

261. $|x| > 2$

$$x > 2 \quad OR \quad x < -2$$

262. $|x| \le 3$

$$x \le 3 \quad AND \quad x \ge -3$$

263. $|x| < 4$

$$x < 4 \quad AND \quad x > -4$$

264. $|x| \ge 5$

$$x \ge 5 \quad OR \quad x \le -5$$

265. $|x| \le 7$

$$x \le 7 \quad AND \quad x \ge -7$$

266. $|x| \le 1$

$$x \le 1 \quad AND \quad x \ge -1$$

267. $|x| > 12$

$$x > 12 \quad OR \quad x < -12$$

268. $|x| \ge 6$

$$x \ge 6 \quad OR \quad x < -6$$

269. $|x| \ge 8$

$$x \ge 8 \quad OR \quad x < -8$$

270. $|x + 1| < 4$

$$x + 1 < 4 \quad AND \quad x + 1 > -4$$
$$x < 4 - 1 \quad AND \quad x > -4 - 1$$
$$x < 3 \quad AND \quad x > -5$$

271. $|x + 2| > 3$

$$x + 2 > 3 \quad OR \quad x + 2 < -3$$
$$x > 3 - 2 \quad OR \quad x < -3 - 2$$
$$x > 1 \quad OR \quad x < -5$$

272. $|x - 3| \ge 6$

$$x - 3 \ge 6 \quad OR \quad x - 3 \le -6$$
$$x \ge 6 + 3 \quad OR \quad x \le -6 + 3$$
$$x \ge 9 \quad OR \quad x \le -3$$

273. $|x + 4| \le 2$

$$x + 4 \le 2 \quad AND \quad x + 4 \ge -2$$
$$x \le 2 - 4 \quad AND \quad x \ge -2 - 4$$
$$x \le -2 \quad AND \quad x \ge -6$$

274. $|x - 2| \ge 6$

$$x - 2 \geq 6 \quad OR \quad x - 2 \leq -6$$
$$x \geq 6 + 2 \quad OR \quad x \leq -6 + 2$$
$$x \geq 8 \quad OR \quad x \leq -4$$

275. $|x - 7| < 5$

$$x - 7 < 5 \quad AND \quad x - 7 > -5$$
$$x < 5 + 7 \quad AND \quad x > -5 + 7$$
$$x < 12 \quad AND \quad x > 2$$

276. $|x - 5| > 4$

$$x - 5 > 4 \quad OR \quad x - 5 < -4$$
$$x > 4 + 5 \quad OR \quad x < -4 + 5$$
$$x > 9 \quad OR \quad x < 1$$

277. $|x - 7| \geq 12$

$$x - 7 \geq 12 \quad OR \quad x - 7 \leq -12$$
$$x \geq 12 + 7 \quad OR \quad x \leq -12 + 7$$
$$x \geq 19 \quad OR \quad x \leq -5$$

278. $|x + 4| \leq 2$

$$x + 4 \leq 2 \quad AND \quad x + 4 \geq -2$$
$$x \leq 2 - 4 \quad AND \quad x \geq -2 - 4$$
$$x \leq -2 \quad AND \quad x \geq -6$$

279. $|x + 5| \geq 7$

$$x + 5 \geq 7 \quad OR \quad x + 5 \leq -7$$
$$x \geq 7 - 5 \quad OR \quad x \leq -7 - 5$$
$$x \geq 2 \quad OR \quad x \leq -12$$

280. $2|x + 1| \geq 4$

$$\frac{2|x + 1|}{2} \geq \frac{4}{2}$$
$$|x + 1| \geq 2$$
$$x + 1 \geq 2 \quad OR \quad x + 1 \leq -2$$
$$x \geq 2 - 1 \quad OR \quad x \leq -2 - 1$$
$$x \geq 1 \quad OR \quad x \leq -3$$

281. $3|x + 2| \leq 3$

$$\frac{3|x + 2|}{3} \leq \frac{3}{3}$$
$$|x + 2| \leq 1$$
$$x + 2 \leq 1 \quad AND \quad x + 2 \geq -1$$
$$x \leq 1 - 2 \quad AND \quad x \geq -1 - 2$$
$$x \leq -1 \quad AND \quad x \geq -3$$

282. $2|x - 3| < 6$

$$\frac{2|x-3|}{2} < \frac{6}{2}$$

$$|x-3| < 3$$

$$x-3 < 3 \quad AND \quad x-3 > -3$$

$$x < 3+3 \quad AND \quad x > -3+3$$

$$x < 6 \quad AND \quad x > 0$$

283. $4|x+7| \geq 12$

$$\frac{4|x+7|}{4} \geq \frac{12}{4}$$

$$|x+7| \geq 3$$

$$x+7 \geq 3 \quad OR \quad x+7 \leq -3$$

$$x \geq 3-7 \quad OR \quad x \leq -3-7$$

$$x \geq -4 \quad OR \quad x \leq -10$$

284. $|x+8|2 < 8$

$$\frac{|x+8|2}{2} < \frac{8}{2}$$

$$|x+8| < 4$$

$$x+8 < 4 \quad AND \quad x+8 > -4$$

$$x < 4-8 \quad AND \quad x > -4-8$$

$$x < -4 \quad AND \quad x > -12$$

285. $6 \geq 3|x-3|$

$$\frac{6}{3} \geq \frac{3|x-3|}{3}$$

$$2 \geq |x-3|$$

$$x-3 \leq 2 \quad AND \quad x-3 \geq -2$$

$$x \leq 2+3 \quad AND \quad x \geq -2+3$$

$$x \leq 5 \quad AND \quad x \geq 1$$

286. $2|x+1| > 1$

$$\frac{2|x+1|}{2} > \frac{1}{2}$$

$$|x+1| > \frac{1}{2}$$

$$x+1 > \frac{1}{2} \quad OR \quad x+1 < -\frac{1}{2}$$

$$x > \frac{1}{2}-1 \quad OR \quad x < -\frac{1}{2}-1$$

$$x > -\frac{1}{2} \quad OR \quad x < -\frac{3}{2}$$

287. $6|x+3| \geq 18$

$$\frac{6|x+3|}{6} \geq \frac{18}{6}$$

$$|x + 3| \geq 3$$
$$x + 3 \geq 3 \quad OR \quad x + 3 \leq -3$$
$$x \geq 3 - 3 \quad OR \quad x \leq -3 - 3$$
$$x \geq 0 \quad OR \quad x \leq -6$$

288. $8 \geq 4|x - 3|$

$$\frac{8}{4} \geq \frac{4|x - 3|}{4}$$
$$2 \geq |x - 3|$$
$$x - 3 \leq 2 \quad AND \quad x - 3 \geq -2$$
$$x \leq 2 + 3 \quad AND \quad x \geq -2 + 3$$
$$x \leq 5 \quad AND \quad x \geq 1$$

289. $12 \geq 2|x + 1|$

$$\frac{12}{2} \geq \frac{2|x + 1|}{2}$$
$$6 \geq |x + 1|$$
$$x + 1 \leq 6 \quad AND \quad x + 1 \geq -6$$
$$x \leq 6 - 1 \quad AND \quad x \geq -6 - 1$$
$$x \leq 5 \quad AND \quad x \geq -7$$

290. $4|x + 2| \geq 6$

$$\frac{4|x + 2|}{4} \geq \frac{6}{4}$$
$$|x + 2| \geq \frac{3}{2}$$
$$x + 2 \geq \frac{3}{2} \quad OR \quad x + 2 \leq -\frac{3}{2}$$
$$x \geq \frac{3}{2} - 2 \quad OR \quad x \leq -\frac{3}{2} - 2$$
$$x \geq -\frac{1}{2} \quad OR \quad x \leq -3\frac{1}{2}$$

Answers to Chapter 5 Linear Equations

291. $y = 3$

292. $y = 1.5$

293. $x = 2$

294. $y = -3$

295. $y = x$

Convert the below equations into y-intercept form:

296. $2y = 4x + 6$

$$\frac{2y}{2} = \frac{4x}{2} + \frac{6}{2}$$

$$y = 2x + 3$$

297. $4y = 4x + 12$

$$\frac{4y}{4} = \frac{4x}{4} + \frac{12}{4}$$

$$y = x + 3$$

298. $6y = 12x + 24$

$$\frac{6y}{6} = \frac{12x}{6} + \frac{24}{6}$$

$$y = 2x + 4$$

299. $5y = 15x + 25$

$$\frac{5y}{5} = \frac{15x}{5} + \frac{25}{5}$$

$$y = 3x + 5$$

300. $7y = 14x + 28$

$$\frac{7y}{7} = \frac{14x}{7} + \frac{28}{7}$$

$$y = 2x + 4$$

301. $10y = 20x + 40$

$$\frac{10y}{10} = \frac{20x}{10} + \frac{40}{10}$$

$$y = 2x + 4$$

302. $12y = -48x + 144$

$$\frac{12y}{12} = -\frac{48x}{12} + \frac{144}{12}$$

$$y = -4x + 12$$

303. $9y = 27x + 45$

$$\frac{9y}{9} = \frac{27x}{9} + \frac{45}{9}$$

$$y = 3x + 5$$

304. $11y = -44x + 121$

$$\frac{11y}{11} = \frac{-44x}{11} + \frac{121}{11}$$

$$y = -4x + 11$$

305. $8y = 24x + 40$

$$\frac{8y}{8} = \frac{24x}{8} + \frac{40}{8}$$

$$y = 3x + 5$$

306. $2y = x + 2$

$$\frac{2y}{2} = \frac{x}{2} + \frac{2}{2}$$

$$y = \frac{x}{2} + 1$$

307. $10y = 10x + 5$

$$\frac{10y}{10} = \frac{10x}{10} + \frac{5}{10}$$

$$y = x + \frac{1}{2}$$

308. $-3y = 6x - 5$

$$\frac{-3y}{-3} = \frac{6x}{-3} - \frac{5}{-3}$$

$$y = -2x + \frac{5}{3}$$

309. $12y = -3x + 6$

$$\frac{12y}{12} = \frac{-3x}{12} + \frac{6}{12}$$

$$y = -\frac{x}{4} + \frac{1}{2}$$

310. $14y = 7x + 28$

$$\frac{14y}{14} = \frac{7x}{14} + \frac{28}{14}$$

$$y = \frac{x}{2} + 2$$

311. $30y = 10x - 15$

$$\frac{30y}{30} = \frac{10x}{30} - \frac{15}{30}$$

$$y = \frac{x}{3} - \frac{1}{2}$$

312. $-17y = 8x - 18$

$$\frac{-17y}{-17} = \frac{8x}{-17} - \frac{18}{-17}$$

$$y = -\frac{8x}{17} + \frac{18}{17}$$

313. $25y = -75x + 100$

$$\frac{25y}{25} = -\frac{75x}{25} + \frac{100}{25}$$

$$y = -3x + 4$$

314. $7y = 28x - 63$

$$\frac{7y}{7} = \frac{28x}{7} - \frac{63}{7}$$

$$y = 4x - 9$$

315. $9y = 54x + 81$

$$\frac{9y}{9} = \frac{54x}{9} + \frac{81}{9}$$

$$y = 6x + 9$$

316. $-2y + 4x = 12$

$$\frac{-2y}{-2} + \frac{4x}{-2} = \frac{12}{-2}$$

$$y - 2x = -6$$

$$y = 2x - 6$$

317. $4y - 16x = 20$

$$\frac{4y}{4} - \frac{16x}{4} = \frac{20}{4}$$

$$y - 4x = 5$$

$$y = 4x + 5$$

318. $4y + 16x = 36$

$$\frac{4y}{4} + \frac{16x}{4} = \frac{36}{4}$$

$$y + 4x = 9$$

$$y = -4x + 9$$

319. $-25y + 5x = 35$

$$\frac{-25y}{-25} + \frac{5x}{-25} = \frac{35}{-25}$$

$$y - \frac{x}{5} = -\frac{7}{5}$$

$$y = \frac{x}{5} - \frac{7}{5}$$

320. $7y = 14x + 28$

$$\frac{7y}{7} = \frac{14x}{7} + \frac{28}{7}$$

$$y = 2x + 4$$

321. $11y - 22x = -66$

$$\frac{11y}{11} - \frac{22x}{11} = -\frac{66}{11}$$

$$y - 2x = -6$$

$$y = 2x - 6$$

322. $12y + 4x = -156$

$$\frac{12y}{12} + \frac{4x}{12} = -\frac{156}{12}$$

$$y + \frac{x}{3} = -13$$

$$y = -\frac{x}{3} - 13$$

323. $2y + 24x = 4$

$$\frac{2y}{2} + \frac{24x}{2} = \frac{4}{2}$$

$$y + 12x = 2$$

$$y = -12x + 2$$

324. $-10y + 55x = 135$

$$\frac{-10y}{-10} + \frac{55x}{-10} = \frac{135}{-10}$$

$$y - \frac{11}{2}x = -\frac{27}{2}$$

$$y = \frac{11}{2}x - \frac{27}{2}$$

325. $8y + 8x = 24$

$$\frac{8y}{8} + \frac{8x}{8} = \frac{24}{8}$$

$$y + x = 3$$

$$y = -x + 3$$

326. $-4y - x = 2$

$$\frac{-4y}{-4} - \frac{x}{-4} = \frac{2}{-4}$$

$$y + \frac{x}{4} = -\frac{1}{2}$$

$$y = -\frac{x}{4} - \frac{1}{2}$$

327. $9y + 3x = 6$

$$\frac{9y}{9} + \frac{3x}{9} = \frac{6}{9}$$

$$y + \frac{x}{3} = \frac{2}{3}$$

$$y = -\frac{x}{3} + \frac{2}{3}$$

328. $-6y + 9x = 12$

$$-6y = -9x + 12$$

$$\frac{-6y}{-6} = \frac{-9x}{-6} + \frac{12}{-6}$$

$$y = \frac{3x}{2} - 2$$

329. $2y - 4x = 16$

$$\frac{2y}{2} - \frac{4x}{2} = \frac{16}{2}$$

$$y - 2x = 8$$

$$y = 2x + 8$$

330. $7y - 21x = 49$

$$\frac{7y}{7} - \frac{21x}{7} = \frac{49}{7}$$

$$y - 3x = 7$$

$$y = 3x + 7$$

331. $15y + 5x = -10$

$$\frac{15y}{15} + \frac{5x}{15} = -\frac{10}{15}$$

$$y + \frac{x}{3} = -\frac{2}{3}$$

$$y = -\frac{x}{3} - \frac{2}{3}$$

332. $-7y + 21x = 84$

$$\frac{-7y}{-7} + \frac{21x}{-7} = \frac{84}{-7}$$

$$y - 3x = -12$$

$$y = 3x - 12$$

333. $5y + 15x = -50$

$$\frac{5y}{5} + \frac{15x}{5} = -\frac{50}{5}$$

$$y + 3x = -10$$

$$y = -3x - 10$$

334. $-13y + 39x = 52$

$$\frac{-13y}{-13} + \frac{39x}{-13} = \frac{52}{-13}$$

$$y - 3x = -4$$

$$y = 3x - 4$$

335. $9y + 18x = -51$

$$\frac{9y}{9} + \frac{18x}{9} = -\frac{51}{9}$$

$$y + 2x = -\frac{17}{3}$$

$$y = -2x - \frac{17}{3}$$

336. $3y = 5x - 45$

$$\frac{3y}{3} = \frac{5x}{3} - \frac{45}{3}$$

$$y = \frac{5x}{3} - 15$$

337. $7y = 3x - 91$

$$\frac{7y}{7} = \frac{3x}{7} - \frac{91}{7}$$

$$y = \frac{3x}{7} - 13$$

338. $8y = -10x + 16$

$$\frac{8y}{8} = \frac{-10x}{8} + \frac{16}{8}$$

$$y = -\frac{5x}{4} + 2$$

339. $-13y = 12x + 26$

$$\frac{-13y}{-13} = \frac{12x}{-13} + \frac{26}{-13}$$

$$y = -\frac{12x}{13} - 2$$

340. $-2y = 6x - 2$

$$\frac{-2y}{-2} = \frac{6x}{-2} - \frac{2}{-2}$$

$$y = -3x + 1$$

341. $-9y = -2x + 72$

$$\frac{-9y}{-9} = \frac{-2x}{-9} + \frac{72}{-9}$$

$$y = \frac{2x}{9} - 8$$

342. $8y = -6x - 68$

$$\frac{8y}{8} = \frac{-6x}{8} - \frac{68}{8}$$

$$y = -\frac{3x}{8} - \frac{68}{8}$$

$$y = -\frac{3x}{4} - \frac{17}{2}$$

343. $5y = 7x + 60$

$$\frac{5y}{5} = \frac{7x}{5} + \frac{60}{5}$$

$$y = \frac{7x}{5} + 12$$

344. $13y = 14x + 169$

$$\frac{13y}{13} = \frac{14x}{13} + \frac{169}{13}$$

$$y = \frac{14x}{13} + 13$$

345. $-8y = 64x + 28$

$$\frac{-8y}{-8} = \frac{64x}{-8} + \frac{28}{-8}$$

$$y = -8x - \frac{7}{2}$$

346. $12y = 7x + 10$

$$\frac{12y}{12} = \frac{7x}{12} + \frac{10}{12}$$

$$y = \frac{7x}{12} + \frac{5}{6}$$

347. $11y = x - 132$

$$\frac{11y}{11} = \frac{x}{11} - \frac{132}{11}$$

$$y = \frac{x}{11} - 12$$

348. $7y = -x - 7$

$$\frac{7y}{7} = \frac{-x}{7} - \frac{7}{7}$$

$$y = -\frac{x}{7} - 1$$

349. $-18y = -18x + 28$

$$\frac{-18y}{-18} = \frac{-18x}{-18} + \frac{28}{-18}$$

$$y = x - \frac{14}{9}$$

350. $-13y = 6x + 14$

$$\frac{-13y}{-13} = \frac{6x}{-13} + \frac{14}{-13}$$

$$y = -\frac{6x}{13} - \frac{14}{13}$$

351. $3y = -8x - 16$

$$\frac{3y}{3} = \frac{-8x}{3} - \frac{16}{3}$$

$$y = -\frac{8x}{3} - \frac{16}{3}$$

352. $7y = 21x + 25$

$$\frac{7y}{7} = \frac{21x}{7} + \frac{25}{7}$$

$$y = 3x + \frac{25}{7}$$

353. $9y = -72x + 81$

$$\frac{9y}{9} = \frac{-72x}{9} + \frac{81}{9}$$

$$y = -8x + 9$$

354. $-5y = -25x + 55$

$$\frac{-5y}{-5} = \frac{-25x}{-5} + \frac{55}{-5}$$

$$y = 5x - 11$$

355. $-12y = -14x - 60$

$$\frac{-12y}{-12} = \frac{-14x}{-12} - \frac{60}{-12}$$

$$y = \frac{7x}{6} + 5$$

356. $15y + 13x = -14$

$$\frac{15y}{15} + \frac{13x}{15} = -\frac{14}{15}$$

$$y = -\frac{13x}{15} - \frac{14}{15}$$

357. $4y - 16x = -32$

$$\frac{4y}{4} - \frac{16x}{4} = -\frac{32}{4}$$

$$y - 4x = -8$$

$$y = 4x - 8$$

358. $-13y + 26x = 44$

$$\frac{-13y}{-13} + \frac{26x}{-13} = \frac{44}{-13}$$

$$y - 2x = \frac{44}{-13}$$

$$y = 2x - \frac{44}{13}$$

359. $22y - 44x = 33$

$$\frac{22y}{22} - \frac{44x}{22} = \frac{33}{22}$$

$$y - 2x = \frac{3}{2}$$

$$y = 2x + \frac{3}{2}$$

360. $-18y = -45x + 36$

$$\frac{-18y}{-18} = \frac{-45x}{-18} + \frac{36}{-18}$$

$$y = \frac{5x}{2} - 2$$

361. $14y + 28x = 56$

$$\frac{14y}{14} + \frac{28x}{14} = \frac{56}{14}$$

$$y + 2x = 4$$

$$y = -2x + 4$$

362. $6y + 12x = 8$

$$\frac{6y}{6} + \frac{12x}{6} = \frac{8}{6}$$

$$y + 2x = \frac{4}{3}$$

$$y = -2x + \frac{4}{3}$$

363. $-20y - 20x = -40$

$$\frac{-20y}{-20} - \frac{20x}{-20} = \frac{-40}{-20}$$

$$y + x = 2$$

$$y = -x + 2$$

364. $25y + 50x = -125$

$$\frac{25y}{25} + \frac{50x}{25} = \frac{-125}{25}$$

$$y + 2x = -5$$

$$y = -2x - 5$$

365. $-6y - 18x = 16$

$$\frac{-6y}{-6} - \frac{18x}{-6} = \frac{16}{-6}$$

$$y + 3x = -\frac{8}{3}$$

$$y = -3x - \frac{8}{3}$$

366. $16y + 32x = 14$

$$\frac{16y}{16} + \frac{32x}{16} = \frac{14}{16}$$

$$y + 2x = \frac{7}{8}$$

$$y = -2x + \frac{7}{8}$$

367. $24y + 48x = 35$

$$\frac{24y}{24} + \frac{48x}{24} = \frac{35}{24}$$

$$y + 2x = \frac{35}{24}$$

$$y = -2x + \frac{35}{24}$$

368. $9y + 45x = 34$

$$\frac{9y}{9} + \frac{45x}{9} = \frac{34}{9}$$

$$y + 5x = \frac{34}{9}$$

$$y = -5x + \frac{34}{9}$$

369. $-4y + 44x = 24$

$$\frac{-4y}{-4} + \frac{44x}{-4} = \frac{24}{-4}$$

$$y - 11x = -6$$

$$y = 11x - 6$$

370. $-8y + 28x = -98$

$$\frac{-8y}{-8} + \frac{28x}{-8} = \frac{-98}{-8}$$

$$y - \frac{7x}{2} = \frac{49}{4}$$

$$y = \frac{7x}{2} + \frac{49}{4}$$

371. $7y + 12x = 120$

$$\frac{7y}{7} + \frac{12x}{7} = \frac{120}{7}$$

$$y + \frac{12x}{7} = \frac{120}{7}$$

$$y = -\frac{12x}{7} + \frac{120}{7}$$

372. $7y + 17x = 6$

$$\frac{7y}{7} + \frac{17x}{7} = \frac{6}{7}$$

$$y + \frac{17x}{7} = \frac{6}{7}$$

$$y = -\frac{17x}{7} + \frac{6}{7}$$

373. $-14y - 16x = 12$

$$\frac{-14y}{-14} - \frac{16x}{-14} = \frac{12}{-14}$$

$$y + \frac{8x}{7} = -\frac{6}{7}$$

$$y = -\frac{8x}{7} - \frac{6}{7}$$

374. $4y - 42x = -4$

$$\frac{4y}{4} - \frac{42x}{4} = \frac{-4}{4}$$

$$y - \frac{21x}{2} = -1$$

$$y = \frac{21x}{2} - 1$$

375. $10y - 8x = 7$

$$\frac{10y}{10} - \frac{8x}{10} = \frac{7}{10}$$

$$y - \frac{4x}{5} = \frac{7}{10}$$

$$y = \frac{4x}{5} + \frac{7}{10}$$

What is the slope of a line that connects the two points?

376. $(1, 1)$ and $(5, 1)$

$$Slope = \frac{y_2 - y_1}{x_2 - x_1} = \frac{1-1}{5-1} = \frac{0}{4} = 0$$

377. $(2, 4)$ and $(1, 6)$

$$Slope = \frac{y_2 - y_1}{x_2 - x_1} = \frac{6-4}{1-2} = \frac{2}{-1} = -2$$

378. $(5, 6)$ and $(6, 3)$

$$Slope = \frac{y_2 - y_1}{x_2 - x_1} = \frac{3-6}{6-5} = \frac{-3}{1} = -3$$

379. $(9, 4)$ and $(8, 2)$

$$Slope = \frac{y_2 - y_1}{x_2 - x_1} = \frac{2-4}{8-9} = \frac{-2}{-1} = 2$$

380. $(3, 6)$ and $(10, 7)$

$$Slope = \frac{y_2 - y_1}{x_2 - x_1} = \frac{7-6}{10-3} = \frac{1}{7}$$

381. $(6, 7)$ and $(8, 10)$

$$Slope = \frac{y_2 - y_1}{x_2 - x_1} = \frac{10-7}{8-6} = \frac{3}{2}$$

382. $(4, 2)$ and $(1, 11)$

$$Slope = \frac{y_2 - y_1}{x_2 - x_1} = \frac{11-2}{1-4} = \frac{9}{-3} = -3$$

383. $(1, 8)$ and $(10, 4)$

$$Slope = \frac{y_2 - y_1}{x_2 - x_1} = \frac{4 - 8}{10 - 1} = \frac{-4}{9} = -\frac{4}{9}$$

384. $(3, 7)$ and $(9, 4)$

$$Slope = \frac{y_2 - y_1}{x_2 - x_1} = \frac{4 - 7}{9 - 3} = \frac{-3}{6} = -\frac{1}{2}$$

385. $(5, 10)$ and $(2, 7)$

$$Slope = \frac{y_2 - y_1}{x_2 - x_1} = \frac{7 - 10}{2 - 5} = \frac{-3}{-3} = 1$$

386. $(10, 8)$ and $(7, 2)$

$$Slope = \frac{y_2 - y_1}{x_2 - x_1} = \frac{2 - 8}{7 - 10} = \frac{-6}{-3} = 2$$

387. $(6, 1)$ and $(6, 4)$

$$Slope = \frac{y_2 - y_1}{x_2 - x_1} = \frac{4 - 1}{6 - 6} = \frac{3}{0} = does\ not\ exist$$

388. $(9, 8)$ and $(9, 8)$

$$Slope = \frac{y_2 - y_1}{x_2 - x_1} = \frac{8 - 8}{9 - 9} = \frac{0}{0} = does\ not\ exist$$

389. $(2, 3)$ and $(0, 0)$

$$Slope = \frac{y_2 - y_1}{x_2 - x_1} = \frac{0 - 3}{0 - 2} = \frac{-3}{-2} = \frac{3}{2}$$

390. $(-7, 5)$ and $(8, -1)$

$$Slope = \frac{y_2 - y_1}{x_2 - x_1} = \frac{-1 - 5}{8 - (-7)} = \frac{-6}{15} = -\frac{2}{5}$$

391. $(2, 7)$ and $(-6, 6)$

$$Slope = \frac{y_2 - y_1}{x_2 - x_1} = \frac{6 - 7}{-6 - 2} = \frac{1}{-8} = -\frac{1}{8}$$

392. $(12, -8)$ and $(0, 4)$

$$Slope = \frac{y_2 - y_1}{x_2 - x_1} = \frac{4 - (-8)}{0 - 12} = \frac{12}{-12} = -1$$

393. $(10, 2)$ and $(6, -12)$

$$Slope = \frac{y_2 - y_1}{x_2 - x_1} = \frac{-12 - 2}{6 - 10} = \frac{-14}{-4} = \frac{7}{2}$$

394. $(-7, 10)$ and $(5, 4)$

$$Slope = \frac{y_2 - y_1}{x_2 - x_1} = \frac{4 - 10}{5 - (-7)} = \frac{-6}{12} = -\frac{1}{2}$$

395. $(8, 11)$ and $(4, 6)$

$$Slope = \frac{y_2 - y_1}{x_2 - x_1} = \frac{6 - 11}{4 - 8} = \frac{-5}{-4} = \frac{5}{4}$$

396. $(5, -12)$ and $(11, 8)$

$$Slope = \frac{y_2 - y_1}{x_2 - x_1} = \frac{8 - (-12)}{11 - 5} = \frac{20}{6} = \frac{10}{3}$$

397. (9, 1) and (-12, 11)

$$Slope = \frac{y_2 - y_1}{x_2 - x_1} = \frac{11 - 1}{-12 - 9} = \frac{10}{-21} = -\frac{10}{21}$$

398. (-1, 8) and (-8, 4)

$$Slope = \frac{y_2 - y_1}{x_2 - x_1} = \frac{4 - 8}{-8 - (-1)} = \frac{-4}{-7} = \frac{4}{7}$$

399. (4, -9) and (6, 8)

$$Slope = \frac{y_2 - y_1}{x_2 - x_1} = \frac{8 - (-9)}{6 - 4} = \frac{17}{2}$$

400. (7, 12) and (1, -9)

$$Slope = \frac{y_2 - y_1}{x_2 - x_1} = \frac{-9 - 12}{1 - 7} = \frac{-21}{-6} = \frac{7}{2}$$

What is the y-intercept form of the line described below?

401. A line with a slope of 1 that passes through points (3, 5)

$y = mx + b$

$y = 1x + b$

$5 = 1(3) + b$

$5 = 3 + b$

$5 - 3 = b$

$2 = b$

$y = x + 2$

402. A line with a slope of 0 that passes through points (8, 6)

$y = mx + b$

$y = 0x + b$

$6 = 0(8) + b$

$6 = b$

$y = 6$

403. A line with a slope of 2 that passes through points (-1, 5)

$y = mx + b$

$y = 2x + b$

$5 = 2(-1) + b$

$5 = -2 + b$

$5 + 2 = b$

$7 = b$

$y = 2x + 7$

404. A line with a slope of 5 that passes through points (-3, -4)

$y = mx + b$

$y = 5x + b$

$-4 = 5(-3) + b$

$-4 = -15 + b$

$15 - 4 = b$

$11 = b$

$y = 5x + 11$

405. A line with a slope of 4 that passes through points (4, 4)

$y = mx + b$

$y = 4x + b$

$4 = 4(4) + b$

$4 = 16 + b$

$4 - 16 = b$

$-12 = b$

$y = 4x - 12$

406. A line with a slope of 8 that passes through points (-1, -1)

$y = mx + b$

$$y = 8x + b$$
$$-1 = 8(-1) + b$$
$$-1 = -8 + b$$
$$-1 + 8 = b$$
$$7 = b$$
$$y = 8x + 7$$

407. A line with a slope of -2 that passes through points $(5, 5)$
$$y = mx + b$$
$$y = -2x + b$$
$$5 = -2(5) + b$$
$$5 = -10 + b$$
$$5 + 10 = b$$
$$15 = b$$
$$y = -2x + 15$$

408. A line with a slope of 4 that passes through points $(6, 7)$
$$y = mx + b$$
$$y = 4x + b$$
$$7 = 4(6) + b$$
$$7 = 24 + b$$
$$-24 + 7 = b$$
$$-17 = b$$
$$y = 4x - 17$$

409. A line with a slope of -2 that passes through points $(9, 1)$
$$y = mx + b$$
$$y = -2x + b$$
$$1 = -2(9) + b$$
$$1 = -18 + b$$
$$1 + 18 = b$$
$$19 = b$$
$$y = -2x + 19$$

410. A line with a slope of 10 that passes through points $(-3, -2)$
$$y = mx + b$$
$$y = 10x + b$$
$$-2 = 10(-3) + b$$
$$-2 = -30 + b$$

$$-2 + 30 = b$$
$$28 = b$$
$$y = 10x + 28$$

411. A line with a slope of -3 that passes through points $(11, 10)$
$$y = mx + b$$
$$y = -3x + b$$
$$10 = -3(11) + b$$
$$10 = -33 + b$$
$$10 + 33 = b$$
$$43 = b$$
$$y = -3x + 43$$

412. A line with a slope of 5 that passes through points $(1, 5)$
$$y = mx + b$$
$$y = 5x + b$$
$$5 = 5(1) + b$$
$$5 = 5 + b$$
$$5 - 5 = b$$
$$0 = b$$
$$y = 5x$$

413. A line with a slope of 4 that passes through points $(8, 12)$
$$y = mx + b$$
$$y = 4x + b$$
$$12 = 4(8) + b$$
$$12 = 32 + b$$
$$12 - 32 = b$$
$$-20 = b$$
$$y = 4x - 20$$

414. A line with a slope of -3 that passes through points $(-12, 7)$
$$y = mx + b$$
$$y = -3x + b$$
$$7 = -3(-12) + b$$
$$7 = 36 + b$$
$$7 - 36 = b$$
$$-29 = b$$
$$y = -3x - 29$$

415. A line with a slope of 7 that passes through points (-5, -4)

$$y = mx + b$$
$$y = 7x + b$$
$$-4 = 7(-5) + b$$
$$-4 = -35 + b$$
$$-4 + 35 = b$$
$$31 = b$$
$$y = 7x + 31$$

416. A line with a slope of 2 that passes through points (-6, 6)

$$y = mx + b$$
$$y = 2x + b$$
$$6 = 2(-6) + b$$
$$6 = -12 + b$$
$$6 + 12 = b$$
$$18 = b$$
$$y = 2x + 18$$

417. A line with a slope of -2 that passes through points (1, -1)

$$y = mx + b$$
$$y = -2x + b$$
$$-1 = -2(1) + b$$
$$-1 = -2 + b$$
$$1 + 2 = b$$
$$1 = b$$
$$y = -2x + 1$$

418. A line with a slope of 4 that passes through points (3, 2)

$$y = mx + b$$
$$y = 4x + b$$
$$2 = 4(3) + b$$
$$2 = 12 + b$$
$$2 - 12 = b$$
$$-10 = b$$
$$y = 4x - 10$$

419. A line with a slope of 7 that passes through points (9, 9)

$$y = mx + b$$
$$y = 7x + b$$
$$9 = 7(9) + b$$
$$9 = 63 + b$$
$$9 - 63 = b$$
$$-54 = b$$
$$y = 7x - 54$$

420. A line with a slope of 4 that passes through points (2, 0)

$$y = mx + b$$
$$y = 4x + b$$
$$0 = 4(2) + b$$
$$0 = 8 + b$$
$$0 - 8 = b$$
$$-8 = b$$
$$y = 4x - 8$$

421. A line with a slope of 8 that passes through points (-2, -6)

$$y = mx + b$$
$$y = 8x + b$$
$$-6 = 8(-2) + b$$
$$-6 = -16 + b$$
$$-6 + 16 = b$$
$$10 = b$$
$$y = 8x + 10$$

422. A line with a slope of -4 that passes through points (7, -8)

$$y = mx + b$$
$$y = -4x + b$$
$$-8 = -4(7) + b$$
$$-8 = -28 + b$$
$$-8 + 28 = b$$
$$20 = b$$
$$y = -4x + 20$$

423. A line with a slope of 2 that passes through points (3, 3)

$$y = mx + b$$
$$y = 2x + b$$
$$3 = 2(3) + b$$
$$3 = 6 + b$$

$$3 - 6 = b$$
$$-3 = b$$
$$y = 2x - 3$$

424. A line with a slope of 3 that passes through points $(7, 7)$

$$y = mx + b$$
$$y = 3x + b$$
$$7 = 3(7) + b$$
$$7 = 21 + b$$
$$7 - 21 = b$$
$$-14 = b$$
$$y = 3x - 14$$

425. A line with a slope of 9 that passes through points $(-8, -7)$

$$y = mx + b$$
$$y = 9x + b$$
$$-7 = 9(-8) + b$$
$$-7 = -72 + b$$
$$-7 + 72 = b$$
$$65 = b$$
$$y = 9x + 65$$

426. A line with a slope of 11 that passes through points $(0, 0)$

$$y = mx + b$$
$$y = 11x + b$$
$$0 = 11(0) + b$$
$$0 = 0 + b$$
$$0 = b$$
$$y = 11x$$

427. A line with a slope of 10 that passes through points $(0, -2)$

$$y = mx + b$$

$$y = 10x + b$$
$$-2 = 10(0) + b$$
$$-2 = 0 + b$$
$$-2 = b$$
$$y = 10x - 2$$

428. A line with a slope of -6 that passes through points $(-1, 7)$

$$y = mx + b$$
$$y = -6x + b$$
$$7 = -6(-1) + b$$
$$7 = 6 + b$$
$$7 - 6 = b$$
$$1 = b$$
$$y = -6x + 1$$

429. A line with a slope of 4 that passes through points $(10, 6)$

$$y = mx + b$$
$$y = 4x + b$$
$$6 = 4(10) + b$$
$$6 = 40 + b$$
$$6 - 40 = b$$
$$-34 = b$$
$$y = 4x - 34$$

430. A line with a slope of -3 that passes through points $(1, 0)$

$$y = mx + b$$
$$y = -3x + b$$
$$0 = -3(1) + b$$
$$0 = -3 + b$$
$$3 = b$$
$$y = -3x + 3$$

Answers to Chapter 6 Systems of Linear Equations

Solve the linear equations using substitution method:

431. $2y + 2x = 4$
$y = x + 2$
$\quad 2(x + 2) + 2x = 4$
$\quad 2x + 4 + 2x = 4$
$\quad\quad 4x = 4 - 4$
$\quad\quad 4x = 0$
$\quad\quad \dfrac{4x}{4} = \dfrac{0}{4}$
$\quad\quad x = 0$
$\quad\quad y = 0 + 2$
$\quad\quad y = 2$
$\quad\quad \underline{x = 0, y = 2}$

432. $3y - 3x = 9$
$y = 2x - 1$
$\quad 3(2x - 1) - 3x = 9$
$\quad 6x - 3 - 3x = 9$
$\quad\quad 3x = 9 + 3$
$\quad\quad 3x = 12$
$\quad\quad \dfrac{3x}{3} = \dfrac{12}{3}$
$\quad\quad x = 4$
$\quad\quad y = 2(4) - 1$
$\quad\quad y = 8 - 1 = 7$
$\quad\quad \underline{x = 4, y = 7}$

433. $4y + 2x = 8$
$y = x + 1$
$\quad 4(x + 1) + 2x = 8$
$\quad 4x + 4 + 2x = 8$
$\quad\quad 6x = 8 - 4$
$\quad\quad 6x = 4$
$\quad\quad \dfrac{6x}{6} = \dfrac{4}{6}$

$\quad\quad x = \dfrac{2}{3}$
$\quad\quad y = \dfrac{2}{3} + 1$
$\quad\quad y = 1\dfrac{2}{3}$

$\quad\quad \underline{x = \dfrac{2}{3}, y = 1\dfrac{2}{3}}$

434. $3y - 7x = 5$
$y = x - 5$
$\quad 3(x - 5) - 7x = 5$
$\quad 3x - 15 - 7x = 5$
$\quad\quad -4x = 5 + 15$
$\quad\quad -4x = 20$
$\quad\quad \dfrac{-4x}{-4} = \dfrac{20}{-4}$
$\quad\quad x = -5$
$\quad\quad y = -5 - 5 = -10$
$\quad\quad \underline{x = -5, y = -10}$

435. $6y - 3x = -9$
$y = x + 4$
$\quad 6(x + 4) - 3x = -9$
$\quad 6x + 24 - 3x = -9$
$\quad\quad 3x = -9 - 24$
$\quad\quad 3x = -33$
$\quad\quad \dfrac{3x}{3} = \dfrac{-33}{3}$
$\quad\quad x = -11$
$\quad\quad y = -11 + 4 = -7$
$\quad\quad \underline{x = -11, y = -7}$

436. $7y + 5x = 17$
$y = 3x - 5$

$$7(3x - 5) + 5x = 17$$
$$21x - 35 + 5x = 17$$
$$26x = 17 + 35$$
$$26x = 52$$
$$\frac{26x}{26} = \frac{52}{26}$$
$$x = 2$$
$$\underline{x = 2, y = 1}$$

437. $6y + 3x = 3$
$$y = 2x + 3$$
$$6(2x + 3) + 3x = 3$$
$$12x + 18 + 3x = 3$$
$$15x = 3 - 18$$
$$15x = -15$$
$$\frac{15x}{15} = \frac{-15}{15}$$
$$x = -1$$
$$y = 2(-1) + 3$$
$$y = -2 + 3 = 1$$
$$\underline{x = -1, y = 1}$$

438. $8y - 2x = -3$
$$y = 4x - 6$$
$$8(4x - 6) - 2x = -3$$
$$32x - 48 - 2x = -3$$
$$30x = -3 + 48$$
$$30x = 45$$
$$\frac{30x}{30} = \frac{45}{30}$$
$$x = \frac{3}{2}$$
$$y = 4\left(\frac{3}{2}\right) - 6$$
$$y = 0$$
$$\underline{x = \frac{3}{2}, y = 0}$$

439. $9y + 5x = 14$
$$y = 5x - 4$$
$$9(5x - 4) + 5x = 14$$
$$45x - 36 + 5x = 14$$
$$50x = 14 + 36$$

$$50x = 50$$
$$\frac{50x}{50} = \frac{50}{50}$$
$$x = 1$$
$$y = 5(1) - 4 = 1$$
$$\underline{x = 1, y = 1}$$

440. $7y - 2x = 24$
$$y = 6x - 8$$
$$7(6x - 8) - 2x = 24$$
$$42x - 56 - 2x = 24$$
$$40x = 24 + 56$$
$$40x = 80$$
$$\frac{40x}{40} = \frac{80}{40}$$
$$x = 2$$
$$y = 6(2) - 8 = 4$$
$$\underline{x = 2, y = 4}$$

441. $4y - 3x = 1$
$$y = 2x - 6$$
$$4(2x - 6) - 3x = 1$$
$$8x - 24 - 3x = 1$$
$$5x = 1 + 24$$
$$5x = 25$$
$$\frac{5x}{5} = \frac{25}{5}$$
$$x = 5$$
$$y = 2(5) - 6 = 4$$
$$\underline{x = 5, y = 4}$$

442. $7y + 8x = 1$
$$y = 6x - 7$$
$$7(6x - 7) + 8x = 1$$
$$42x - 49 + 8x = 1$$
$$50x = 1 + 49$$
$$50x = 50$$
$$\frac{50x}{50} = \frac{50}{50}$$
$$x = 1$$
$$y = 6(1) - 7$$
$$y = -1$$
$$\underline{x = 1, y = -1}$$

443. $6y + 4x = -5$

$y = 7x + 3$

$6(7x + 3) + 4x = -5$

$42x + 18 + 4x = -5$

$46x = -5 - 18$

$46x = -23$

$\dfrac{46x}{46} = \dfrac{-23}{46}$

$x = -\dfrac{1}{2}$

$y = 7\left(-\dfrac{1}{2}\right) + 3$

$y = -\dfrac{7}{2} + 3 = -\dfrac{1}{2}$

$\underline{x = -\dfrac{1}{2}, y = -\dfrac{1}{2}}$

444. $9y - 3x = -42$

$y = 7x - 8$

$9(7x - 8) - 3x = -42$

$63x - 72 - 3x = -42$

$60x = -42 + 72$

$60x = 30$

$\dfrac{60x}{60} = \dfrac{30}{60}$

$x = \dfrac{30}{60} = \dfrac{1}{2}$

$y = 7\left(\dfrac{1}{2}\right) - 8$

$y = \dfrac{7}{2} - \dfrac{16}{2} = -\dfrac{9}{2}$

$\underline{x = \dfrac{1}{2}, y = -\dfrac{9}{2}}$

445. $5y + 9x = 16$

$y = x - 8$

$5(x - 8) + 9x = 16$

$5x - 40 + 9x = 16$

$14x = 16 + 40$

$14x = 56$

$\dfrac{14x}{14} = \dfrac{56}{14}$

$x = 4$

$y = 4 - 8$

$y = -4$

$\underline{x = 4, y = -4}$

446. $2y - 2x = 6$

$y = 3x - 3$

$2(3x - 3) - 2x = 6$

$6x - 6 - 2x = 6$

$4x = 6 + 6$

$4x = 12$

$\dfrac{4x}{4} = \dfrac{12}{4}$

$x = 3$

$y = 3(3) - 3 = 6$

$\underline{x = 3, y = 6}$

447. $8y + 9x = 9$

$y = 5x - 5$

$8(5x - 5) + 9x = 9$

$40x - 40 + 9x = 9$

$49x = 49$

$\dfrac{49x}{49} = \dfrac{49}{49}$

$x = 1$

$y = 5(1) - 5 = 0$

$\underline{x = 1, y = 0}$

448. $5y + 5x = 5$

$y = 6x - 6$

$5(6x - 6) + 5x = 5$

$30x - 30 + 5x = 5$

$35x = 5 + 30$

$35x = 35$

$\dfrac{35x}{35} = \dfrac{35}{35}$

$x = 1$

$y = 6(1) - 6 = 0$

$\underline{x = 1, y = 0}$

449. $7y - 6x = -4$

$y = 2x + 4$

$7(2x + 4) - 6x = -4$

$14x + 28 - 6x = -4$

$8x = -4 - 28$

$$8x = -32$$
$$\frac{8x}{8} = \frac{-32}{8}$$
$$x = -4$$
$$y = 2(-4) + 4 = -4$$
$$\underline{x = -4, y = -4}$$

450. $2y + 4x = 6$
$$2y = 6x - 4$$
$$\frac{2y}{2} = \frac{6x}{2} - \frac{4}{2}$$
$$y = 3x - 2$$
$$2(3x - 2) + 4x = 6$$
$$6x - 4 + 4x = 6$$
$$10x = 6 + 4$$
$$10x = 10$$
$$\frac{10x}{10} = \frac{10}{10}$$
$$x = 1$$
$$y = 3(1) - 2$$
$$y = 1$$
$$\underline{x = 1, y = 1}$$

451. $2y + 4x = 6$
$$3y - 3x = 9$$
$$\frac{3y}{3} - \frac{3x}{3} = \frac{9}{3}$$
$$y - x = 3$$
$$y = x + 3$$
$$2(x + 3) + 4x = 6$$
$$2x + 6 + 4x = 6$$
$$6x = 6 - 6$$
$$6x = 0$$
$$x = 0$$
$$y = 0 + 3 = 3$$
$$\underline{x = 0, y = 3}$$

452. $2y + 4x = 6$
$$4y + 4x = 8$$
$$\frac{4y}{4} + \frac{4x}{4} = \frac{8}{4}$$
$$y + x = 2$$
$$y = -x + 2$$

$$2(-x + 2) + 4x = 6$$
$$-2x + 4 + 4x = 6$$
$$2x = 6 - 4$$
$$2x = 2$$
$$\frac{2x}{2} = \frac{2}{2}$$
$$x = 1$$
$$y = -1 + 2 = 1$$
$$\underline{x = 1, y = 1}$$

453. $4y - 2x = 6$
$$3y + 3x = 9$$
$$\frac{3y}{3} + \frac{3x}{3} = \frac{9}{3}$$
$$y + x = 3$$
$$y = -x + 3$$
$$4(-x + 3) - 2x = 6$$
$$-4x + 12 - 2x = 6$$
$$-6x = 6 - 12$$
$$-6x = -6$$
$$\frac{-6x}{-6} = \frac{-6}{-6}$$
$$x = 1$$
$$y = -1 + 3 = 2$$
$$\underline{x = 1, y = 2}$$

454. $2y + 4x = 20$
$$5y + 2x = 10$$
$$\frac{2y}{2} + \frac{2x}{2} = \frac{10}{2}$$
$$y + x = 5$$
$$y = -x + 5$$
$$2(-x + 5) + 4x = 20$$
$$-2x + 10 + 4x = 20$$
$$2x = 20 - 10 = 10$$
$$\frac{2x}{2} = \frac{10}{2}$$
$$x = 5$$
$$y = -5 + 5 = 0$$
$$\underline{x = 5, y = 0}$$

455. $2y - 5x = 10$
$$3y - 6x = 12$$

$$\frac{3y}{3} - \frac{6x}{3} = \frac{12}{3}$$
$$y - 2x = 4$$
$$y = 2x + 4$$
$$2(2x + 4) - 5x = 10$$
$$4x + 8 - 5x = 10$$
$$-x = 10 - 8$$
$$-x = 2$$
$$\frac{-x}{-1} = \frac{2}{-1}$$
$$x = -2$$
$$y = 2(-2) + 4$$
$$y = 0$$
$$\underline{x = -2, y = 0}$$

456. $4y + 6x = 20$
$$2y + 4x = 8$$
$$\frac{2y}{2} + \frac{4x}{2} = \frac{8}{2}$$
$$y + 2x = 4$$
$$y = -2x + 4$$
$$4(-2x + 4) + 6x = 20$$
$$-8x + 16 + 6x = 20$$
$$-2x = 20 - 16$$
$$-2x = 4$$
$$\frac{-2x}{-2} = \frac{4}{-2}$$
$$x = -2$$
$$y + 2(-2) = 4$$
$$y - 4 = 4$$
$$y = 4 + 4 = 8$$
$$\underline{x = -2, y = 8}$$

457. $4y - 8x = 12$
$$2y - 5x = 10$$
$$\frac{4y}{4} - \frac{8x}{4} = \frac{12}{4}$$
$$y - 2x = 3$$
$$y = 2x + 3$$
$$2(2x + 3) - 5x = 10$$
$$4x + 6 - 5x = 10$$
$$-x = 10 - 6$$

$$-x = 4$$
$$\frac{-x}{-1} = \frac{4}{-1}$$
$$x = -4$$
$$y = 2(-4) + 3$$
$$y = -8 + 3 = -5$$
$$\underline{x = -4, y = -5}$$

458. $6y + 6x = 18$
$$4y + 8x = 8$$
$$\frac{6y}{6} + \frac{6x}{6} = \frac{18}{6}$$
$$y + x = 3$$
$$y = -x + 3$$
$$4(-x + 3) + 8x = 8$$
$$-4x + 12 + 8x = 8$$
$$4x = 8 - 12 = -4$$
$$4x = -4$$
$$\frac{4x}{4} = \frac{-4}{4}$$
$$x = -1$$
$$y = -(-1) + 3 = 4$$
$$\underline{x = -1, y = 4}$$

459. $5y - 5x = 25$
$$9y + 6x = 15$$
$$\frac{5y}{5} - \frac{5x}{5} = \frac{25}{5}$$
$$y - x = 5$$
$$y = x + 5$$
$$9(x + 5) + 6x = 15$$
$$9x + 45 + 6x = 15$$
$$15x = 15 - 45$$
$$15x = -30$$
$$\frac{15x}{15} = \frac{-30}{15}$$
$$x = -2$$
$$y = -2 + 5 = 3$$
$$\underline{x = -2, y = 3}$$

460. $7y - 7x = 21$
$$6y - 8x = 24$$

$$\frac{7y}{7} - \frac{7x}{7} = \frac{21}{7}$$
$$y - x = 3$$
$$y = x + 3$$
$$8(x + 3) - 8x = 24$$
$$6x + 18 - 8x = 24$$
$$-2x = 24 - 18$$
$$-2x = 6$$
$$\frac{-2x}{-2} = \frac{6}{-2}$$
$$x = -3$$
$$y = -3 + 3 = 0$$
$$\underline{x = -3, y = 0}$$

461. $8y + 6x = 28$
$$2y + 4x = 12$$
$$\frac{2y}{2} + \frac{4x}{2} = \frac{12}{2}$$
$$y + 2x = 6$$
$$y = -2x + 6$$
$$8(-2x + 6) + 6x = 28$$
$$-16x + 48 + 6x = 28$$
$$-10x = 28 - 48$$
$$-10x = -20$$
$$\frac{-10x}{-10} = \frac{-20}{-10}$$
$$x = 2$$
$$y = -2(2) + 6 = 2$$
$$\underline{x = 2, y = 2}$$

462. $7y - 8x = 54$
$$3y + 6x = 42$$
$$\frac{3y}{3} + \frac{6x}{3} = \frac{42}{3}$$
$$y + 2x = 14$$
$$y = -2x + 14$$
$$7(-2x + 14) - 8x = 54$$
$$-14x + 98 - 8x = 54$$
$$-22x = 54 - 98$$
$$-22x = -44$$
$$\frac{-22x}{-22} = \frac{-44}{-22}$$

$$x = 2$$
$$y = -2(2) + 14$$
$$y = -4 + 14 = 10$$
$$\underline{x = 2, y = 10}$$

463. $6y - 6x = -24$
$$7y + 7x = 28$$
$$\frac{7y}{7} + \frac{7x}{7} = \frac{28}{7}$$
$$y + x = 4$$
$$y = 4 - x$$
$$6(4 - x) - 6x = -24$$
$$24 - 6x - 6x = -24$$
$$-12x = -24 - 24$$
$$-12x = -48$$
$$\frac{-12x}{-12} = \frac{-48}{-12}$$
$$x = 4$$
$$y = 4 - 4 = 0$$
$$\underline{x = 4, y = 0}$$

464. $3y + 3x = 6$
$$8y - 16x = -32$$
$$\frac{3y}{3} + \frac{3x}{3} = \frac{6}{3}$$
$$y + x = 2$$
$$y = 2 - x$$
$$8(2 - x) - 16x = -32$$
$$16 - 8x - 16x = -32$$
$$-24x = -32 - 16$$
$$\frac{-24x}{-24} = \frac{-48}{-24}$$
$$x = 2$$
$$y = 2 - 2 = 0$$
$$\underline{x = 2, y = 0}$$

465. $5y - 15x = 15$
$$3y - 4x = 19$$
$$\frac{5y}{5} - \frac{15x}{5} = \frac{15}{5}$$
$$y - 3x = 3$$
$$y = 3x + 3$$
$$3(3x + 3) - 4x = 19$$

$$9x + 9 - 4x = 19$$
$$5x = 19 - 9$$
$$5x = 10$$
$$\frac{5x}{5} = \frac{10}{5}$$
$$x = 2$$
$$y = 3(2) + 3 = 9$$
$$\underline{x = 2, y = 9}$$

466. $15y + 10x = 5$
$$2y - 2x = 4$$
$$\frac{2y}{2} - \frac{2x}{2} = \frac{4}{2}$$
$$y - x = 2$$
$$y = 2 + x$$
$$15(2 + x) + 10x = 5$$
$$30 + 15x + 10x = 5$$
$$25x = 5 - 30$$
$$25x = -25$$
$$\frac{25x}{25} = \frac{-25}{25}$$
$$x = -1$$
$$y = 2 - 1 = 1$$
$$\underline{x = -1, y = 1}$$

467. $2y - 6x = 4$
$$4y - 10x = 6$$
$$\frac{2y}{2} - \frac{6x}{2} = \frac{4}{2}$$
$$y - 3x = 2$$
$$y = 3x + 2$$
$$4(3x + 2) - 10x = 6$$
$$12x + 8 - 10x = 6$$
$$2x = 6 - 8$$
$$2x = -2$$
$$\frac{2x}{2} = \frac{-2}{2}$$
$$x = -1$$
$$y = 3(-1) + 2 = -1$$
$$\underline{x = -1, y = -1}$$

468. $7y - 14x = 7$

$$8y - 14x = 4$$
$$\frac{7y}{7} - \frac{14x}{7} = \frac{7}{7}$$
$$y - 2x = 1$$
$$y = 2x + 1$$
$$8(2x + 1) - 14x = 4$$
$$16x + 8 - 14x = 4$$
$$2x = 4 - 8$$
$$2x = -4$$
$$\frac{2x}{2} = \frac{-4}{2}$$
$$x = -2$$
$$y = 2(-2) + 1 = -3$$
$$\underline{x = -2, y = -3}$$

469. $4y + 8x = 0$
$$2y - 4x = 8$$
$$\frac{2y}{2} - \frac{4x}{2} = \frac{8}{2}$$
$$y - 2x = 4$$
$$y = 4 + 2x$$
$$4(4 + 2x) + 8x = 0$$
$$16 + 8x + 8x = 0$$
$$16 + 16x = 0$$
$$16x = -16$$
$$\frac{16x}{16} = \frac{-16}{16}$$
$$x = -1$$
$$y = 4 + 2(-1) = 2$$
$$\underline{x = -1, y = 2}$$

470. $4y + 4x = 12$
$$6y + 4x = 4$$
$$\frac{4y}{4} + \frac{4x}{4} = \frac{12}{4}$$
$$y + x = 3$$
$$y = 3 - x$$
$$6(3 - x) + 4x = 4$$
$$18 - 6x + 4x = 4$$
$$-2x = 4 - 18$$
$$-2x = -14$$

$$\frac{-2x}{-2} = \frac{-14}{-2}$$
$$x = 7$$
$$y = 3 - 7 = -4$$
$$\underline{x = 7, y = -4}$$

471. $12y + 3x = 6$

$3y + 6x = 12$

$$\frac{3y}{3} + \frac{6x}{3} = \frac{12}{3}$$
$$y + 2x = 4$$
$$y = -2x + 4$$
$$12(-2x + 4) + 3x = 6$$
$$-24x + 48 + 3x = 6$$
$$-21x = 6 - 48$$
$$-21x = -42$$
$$\frac{-21x}{-21} = \frac{-42}{-21}$$
$$x = 2$$
$$y = -2(2) + 4 = 0$$
$$\underline{x = 2, y = 0}$$

472. $6y + 3x = 9$

$2y + 6x = 18$

$$\frac{2y}{2} + \frac{6x}{2} = \frac{18}{2}$$
$$y + 3x = 9$$
$$y = -3x + 9$$
$$6(-3x + 9) + 3x = 9$$
$$-18x + 54 + 3x = 9$$
$$-15x = 9 - 54$$
$$-15x = -45$$
$$\frac{-15x}{-15} = \frac{-45}{-15}$$
$$x = 3$$
$$y = -3(3) = 9 = 0$$
$$\underline{x = 3, y = 0}$$

473. $14y + 20x = 10$

$3y + 6x = 9$

$$\frac{3y}{3} + \frac{6x}{3} = \frac{9}{3}$$
$$y + 2x = 3$$

$$y = 3 - 2x$$
$$14y + 20x = 10$$
$$14(3 - 2x) + 20x = 10$$
$$42 - 28x + 20x = 10$$
$$-8x = 10 - 42$$
$$-8x = -32$$
$$\frac{-8x}{-8} = \frac{-32}{-8}$$
$$x = 4$$
$$y = 3 - 2(4) = -5$$
$$\underline{x = 4, y = -5}$$

474. $6y - 8x = 12$

$4y - 8x = 16$

$$\frac{4y}{4} - \frac{8x}{4} = \frac{16}{4}$$
$$y - 2x = 4$$
$$y = 4 + 2x$$
$$6(4 + 2x) - 8x = 12$$
$$24 + 12x - 8x = 12$$
$$4x = 12 - 24$$
$$4x = -24$$
$$\frac{4x}{4} = \frac{-12}{4}$$
$$x = -3$$
$$y = 4 + 2(-3) = -2$$
$$\underline{x = -3, y = -2}$$

475. $6y + 9x = 18$

$4y + 8x = 16$

$$\frac{4y}{4} + \frac{8x}{4} = \frac{16}{4}$$
$$y + 2x = 4$$
$$y = -2x + 4$$
$$6(-2x + 4) + 9x = 18$$
$$-12x + 24 + 9x = 18$$
$$-3x = 18 - 24$$
$$-3x = -6$$
$$\frac{-3x}{-3} = \frac{-6}{-3}$$
$$x = 2$$
$$y = -2(2) + 4 = 0$$

$$x = 2, y = 0$$

476. $3y - 6x = 18$

$3y + 3x = 9$

$$\frac{3y}{3} + \frac{3x}{3} = \frac{9}{3}$$

$$y + x = 3$$

$$y = 3 - x$$

$$3(3 - x) - 6x = 18$$

$$9 - 3x - 6x = 18$$

$$-9x = 18 - 9$$

$$-9x = 9$$

$$\frac{-9x}{-9} = \frac{9}{-9}$$

$$x = -1$$

$$y = 3 - (-1) = 4$$

$$\underline{x = -1, y = 4}$$

477. $6y + 18x = 12$

$4y + 2x = 18$

$$\frac{6y}{6} + \frac{18x}{6} = \frac{12}{6}$$

$$y + 3x - 2$$

$$y = 2 - 3x$$

$$4(2 - 3x) + 2x = 18$$

$$8 - 12x + 2x = 18$$

$$-10x = 18 - 8$$

$$-10x = 10$$

$$\frac{-10x}{-10} = \frac{10}{-10}$$

$$x = -1$$

$$y = 2 - 3(-1) = 5$$

$$\underline{x = -1, y = 5}$$

478. $4y - 8x = 8$

$14y + 12x = -12$

$$\frac{4y}{4} - \frac{8x}{4} = \frac{8}{4}$$

$$y - 2x = 2$$

$$y = 2x + 2$$

$$14(2x + 2) + 12x = -12$$

$$28x + 28 + 12x = -12$$

$$40x = -40$$

$$\frac{40x}{40} = \frac{-40}{40}$$

$$x = -1$$

$$y = 2(-1) + 2 = 0$$

$$\underline{x = -1, y = 0}$$

479. $9y - 6x = 18$

$5y - 5x = 25$

$$\frac{5y}{5} - \frac{5x}{5} = \frac{25}{5}$$

$$y - x = 5$$

$$y = x + 5$$

$$9(x + 5) - 6x = 18$$

$$9x + 45 - 6x = 18$$

$$3x = 18 - 45$$

$$3x = -27$$

$$\frac{3x}{3} = \frac{-27}{3}$$

$$x = -9$$

$$y = -9 + 5 = -4$$

$$\underline{x = -9, y = -4}$$

480. $8y + 16x = 24$

$4y + 6x = 4$

$$\frac{8y}{8} + \frac{16x}{8} = \frac{24}{8}$$

$$y + 2x = 3$$

$$y = -2x + 3$$

$$4(-2x + 3) + 6x = 4$$

$$-8x + 12 + 6x = 4$$

$$-2x = 4 - 12$$

$$\frac{-2x}{-2} = \frac{-8}{-2}$$

$$x = 4$$

$$y = -2(4) + 3 = -5$$

$$\underline{x = 4, y = -5}$$

481. $5y - 10x = 25$

$4y + 3x = 9$

$$\frac{5y}{5} - \frac{10x}{5} = \frac{25}{5}$$

$$y - 2x = 5$$

$$y = 5 + 2x$$

$$4(5 + 2x) + 3x = 9$$
$$20 + 8x + 3x = 9$$
$$11x = 9 - 20$$
$$11x = -11$$
$$\frac{11x}{11} = \frac{-11}{11}$$
$$x = -1$$
$$y = 5 + 2(-1) = 3$$
$$\underline{x = -1, y = 3}$$

482. $7y - 8x = 4$
$$14y - 14x = 7$$
$$\frac{14y}{14} - \frac{14x}{14} = \frac{7}{14}$$
$$y - x = \frac{1}{2}$$
$$y = x + \frac{1}{2}$$
$$7\left(x + \frac{1}{2}\right) - 8x = 4$$
$$7x + \frac{7}{2} - 8x = 4$$
$$-x = 4 - \frac{7}{2}$$
$$-x = \frac{8}{2} - \frac{7}{2}$$
$$-x = \frac{1}{2}$$
$$\frac{-x}{-1} = \frac{\frac{1}{2}}{-1}$$
$$x = -\frac{1}{2}$$
$$y = -\frac{1}{2} + \frac{1}{2} = 0$$
$$\underline{x = -\frac{1}{2}, y = 0}$$

483. $6y + 4x = 34$
$$8y + 40x = -24$$
$$\frac{8y}{8} + \frac{40x}{8} = \frac{24}{8}$$
$$y + 5x = -3$$
$$y = -5x - 3$$

$$6(-5x - 3) + 4x = 34$$
$$-30x - 18 + 4x = 34$$
$$-26x = 52$$
$$\frac{-26x}{-26} = \frac{52}{-26}$$
$$x = -2$$
$$y = -5(-2) - 3 = 7$$
$$\underline{x = -2, y = 7}$$

484. $9y + 27x = 72$
$$6y + 12x = 18$$
$$\frac{9y}{9} + \frac{27x}{9} = \frac{72}{9}$$
$$y + 3x = 8$$
$$y = -3x + 8$$
$$6(-3x + 8) + 12x = 18$$
$$-18x + 48 + 12x = 18$$
$$-6x = 18 - 48$$
$$-6x = -30$$
$$\frac{-6x}{-6} = \frac{-30}{-6}$$
$$x = 5$$
$$y = -3(5) + 8 = -7$$
$$\underline{x = 5, y = -7}$$

485. $9y + 36x = 81$
$$5y + 5x = 60$$
$$\frac{5y}{5} + \frac{5x}{5} = \frac{60}{5}$$
$$y + x = 12$$
$$y = -x + 12$$
$$9(-x + 12) + 36x = 81$$
$$-9x + 108 + 36x = 81$$
$$27x = 81 - 108$$
$$\frac{27x}{27} = \frac{-27}{27}$$
$$x = -1$$
$$y = -(-1) + 12 = 13$$
$$\underline{x = -1, y = 13}$$

486. $4y + 20x = 40$
$$9y + 27x = 54$$

$$\frac{4y}{4} + \frac{20x}{4} = \frac{40}{4}$$

$$y + 5x = 10$$

$$y = -5x + 10$$

$$9(-5x + 10) + 27x = 54$$

$$-45x + 90 + 27x = 54$$

$$-18x = 54 - 90$$

$$-18x = -36$$

$$\frac{-18x}{-18} = \frac{-36}{-18}$$

$$x = 2$$

$$y = -5(2) + 10 = 0$$

$$\underline{x = 2, y = 0}$$

487. $18y + 120x = 768$

$$3y + 60x = 48$$

$$\frac{3y}{3} + \frac{60x}{3} = \frac{48}{3}$$

$$y + 20x = 16$$

$$y = -20x + 16$$

$$18(-20x + 16) + 120x = 768$$

$$-360x + 288 + 120x = 768$$

$$-240x = 480$$

$$\frac{-240x}{-240} = \frac{480}{-240}$$

$$x = -2$$

$$y = -20(-2) + 16 = 56$$

$$\underline{x = -2, y = 56}$$

488. $25y + 50x = 75$

$$60y + 20x = -20$$

$$\frac{25y}{25} + \frac{50x}{25} = \frac{75}{25}$$

$$y + 2x = 3$$

$$y = -2x + 3$$

$$60(-2x + 3) + 20x = -20$$

$$-120x + 180 + 20x = -20$$

$$-100x = -20 - 180$$

$$-100x = -200$$

$$\frac{-100x}{-100} = \frac{-200}{-100}$$

$$x = 2$$

$$y = -2(2) + 3 = -1$$

$$\underline{x = 2, y = -1}$$

489. $14y - 21x = -7$

$$5y - 15x = 20$$

$$\frac{5y}{5} - \frac{15x}{5} = \frac{20}{5}$$

$$y - 3x = 4$$

$$y = 3x + 4$$

$$14(3x + 4) - 21x = -7$$

$$42x + 56 - 21x = -7$$

$$21x = -63$$

$$\frac{21x}{21} = \frac{-63}{21}$$

$$x = -3$$

$$y = 3(-3) + 4 = -5$$

$$\underline{x = -3, y = -5}$$

490. $7y + 13x = 17$

$$8y - 8x = 8$$

$$\frac{8y}{8} - \frac{8x}{8} = \frac{8}{8}$$

$$y - x = 1$$

$$y = x + 1$$

$$7(x + 1) + 13x = 1$$

$$7x + 7 + 13x = 17$$

$$20x = 17 - 7$$

$$20x = 10$$

$$\frac{20x}{20} = \frac{10}{20}$$

$$x = \frac{1}{2}$$

$$y = \frac{1}{2} + 1 = \frac{3}{2}$$

$$\underline{x = \frac{1}{2}, y = \frac{3}{2}}$$

491. $4y + 16x = 8$

$$6y - 12x = 12$$

$$\frac{4y}{4} + \frac{16x}{4} = \frac{8}{4}$$

$$y + 4x = 2$$

$$y = 2 - 4x$$

$$6(2 - 4x) - 12x = 12$$

$$12 - 24x - 12x = 12$$
$$-36x = 12 - 12$$
$$-36x = 0$$
$$\frac{-36x}{-36} = \frac{0}{36}$$
$$x = 0$$
$$y = 2 - 4(0) = 2$$
$$\underline{x = 0, y = 2}$$

492. $9y - 4x = 13$
$$12y - 48x = 60$$
$$\frac{12y}{12} - \frac{48x}{12} = \frac{60}{12}$$
$$y - 4x = 5$$
$$y = 4x + 5$$
$$9(4x + 5) - 4x = 13$$
$$36x + 45 - 4x = 13$$
$$32x = 13 - 45$$
$$32x = -32$$
$$\frac{32x}{32} = \frac{-32}{32}$$
$$x = -1$$
$$y = 4(-1) + 5 = 1$$
$$\underline{x = -1, y = 1}$$

493. $14y + 6x = 86$
$$5y - 40x = -180$$
$$\frac{5y}{5} - \frac{40x}{5} = \frac{-180}{5}$$
$$y - 8x = -36$$
$$y = -36 + 8x$$
$$14(-36 + 8x) + 6x = 86$$
$$-504 + 112x + 6x = 86$$
$$118x = 590$$
$$\frac{118x}{118} = \frac{590}{118}$$
$$x = 5$$
$$y = -36 + 8(5) = 4$$
$$\underline{x = 5, y = 4}$$

494. $23y + 12x = 81$
$$17y - 34x = 17$$

$$\frac{17y}{17} - \frac{34x}{17} = \frac{17}{17}$$
$$y - 2x = 1$$
$$y = 1 + 2x$$
$$23(1 + 2x) + 12x = 81$$
$$23 + 46x + 12x = 81$$
$$58x = 81 - 23$$
$$58x = 58$$
$$\frac{58x}{58} = \frac{58}{58}$$
$$x = 1$$
$$y = 1 + 2(1) = 3$$
$$\underline{x = 1, y = 3}$$

495. $9y + 18x = 18$
$$16y + 9x = 9$$
$$\frac{9y}{9} + \frac{18x}{9} = \frac{18}{9}$$
$$y + 2x = 2$$
$$y = -2x + 2$$
$$16(-2x + 2) + 9x = 9$$
$$-32x + 32 + 9x = 9$$
$$-23x = 9 - 32$$
$$-23x = -23$$
$$\frac{-23x}{-23} = \frac{-23}{-23}$$
$$x = 1$$
$$y = -2(1) + 2 = 0$$
$$\underline{x = 1, y = 0}$$

496. $5y - 15x = 25$
$$7y - 6x = 5$$
$$\frac{5y}{5} - \frac{15x}{5} = \frac{25}{5}$$
$$y - 3x = 5$$
$$y = 3x + 5$$
$$7(3x + 5) - 6x = 5$$
$$21x + 35 - 6x = 5$$
$$15x = 5 - 35$$
$$15x = -30$$
$$\frac{15x}{15} = \frac{-30}{15}$$

$$x = -2$$
$$y = 3(-2) + 5 = -1$$
$$\underline{x = -2, y = -1}$$

497. $6y + 7x = -10$
$9y + 27x = 18$
$$\frac{9y}{9} + \frac{27x}{9} = \frac{18}{9}$$
$$y + 3x = 2$$
$$y = -3x + 2$$
$$6(-3x + 2) + 7x = -10$$
$$-18x + 12 + 7x = -10$$
$$-11x = -10 - 12$$
$$-11x = -22$$
$$\frac{-11x}{-11} = \frac{-22}{-11}$$
$$x = 2$$
$$y = -3(2) + 2 = -4$$
$$\underline{x = 2, y = -4}$$

498. $2y + 12x = 16$
$8y - 9x = 7$
$$\frac{2y}{2} + \frac{12x}{2} = \frac{16}{2}$$
$$y + 6x = 8$$
$$y = 8 - 6x$$
$$8(8 - 6x) - 9x = 7$$
$$64 - 48x - 9x = 7$$
$$-57x = 7 - 64$$
$$-57x = -57$$
$$\frac{-57x}{-57} = \frac{-57}{-57}$$
$$x = 1$$
$$y = 8 - 6(1) = 2$$
$$\underline{x = 1, y = 2}$$

499. $5y - 25x = 15$
$3y + 15x = 39$
$$\frac{5y}{5} - \frac{25x}{5} = \frac{15}{5}$$
$$y - 5x = 3$$
$$y = 3 + 5x$$
$$3(3 + 5x) + 15x = 39$$

$$9 + 15x + 15x = 39$$
$$30x = 39 - 9$$
$$30x = 30$$
$$\frac{30x}{30} = \frac{30}{30}$$
$$x = 1$$
$$y = 3 + 5(1) = 8$$
$$\underline{x = 1, y = 8}$$

500. $12y + 12x = 12$
$14y + 9x = 9$
$$12y + 12x = 12$$
$$\frac{12y}{12} + \frac{12x}{12} = \frac{12}{12}$$
$$y + x = 1$$
$$y = -x + 1$$
$$14(-x + 1) + 9x = 9$$
$$-14x + 14 + 9x = 9$$
$$-5x = 9 - 14$$
$$-5x = -5$$
$$\frac{-5x}{-5} = \frac{-5}{-5}$$
$$x = 1$$
$$y = -1 + 1 = 0$$
$$\underline{x = 1, y = 0}$$

501. $16y + 8x = 8$
$18y + 2x = 23$
$$\frac{16y}{16} + \frac{8x}{16} = \frac{8}{16}$$
$$y + \frac{x}{2} = \frac{1}{2}$$
$$y = -\frac{x}{2} + \frac{1}{2}$$
$$18\left(-\frac{x}{2} + \frac{1}{2}\right) + 2x = 23$$
$$-\frac{18x}{2} + \frac{18}{2} + 2x = 23$$
$$-9x + 9 + 2x = 23$$
$$-7x = 23 - 9$$
$$-7x = 14$$
$$\frac{-7x}{-7} = \frac{14}{-7}$$

$$x = -2$$

$$y = -\frac{-2}{2} + \frac{1}{2} = \frac{3}{2}$$

$$\underline{x = -2, y = \frac{3}{2}}$$

502. $14y - 9x = 29$

$$10y - 5x = 15$$

$$\frac{10y}{10} - \frac{5x}{10} = \frac{15}{10}$$

$$y - \frac{x}{2} = \frac{3}{2}$$

$$y = \frac{x}{2} + \frac{3}{2}$$

$$14y - 9x = 29$$

$$14\left(\frac{x}{2} + \frac{3}{2}\right) - 9x = 29$$

$$\frac{14x}{2} + \frac{42}{2} - 9x = 29$$

$$7x + 21 - 9x = 29$$

$$-2x = 29 - 21$$

$$-2x = 8$$

$$\frac{-2x}{-2} = \frac{8}{-2}$$

$$x = -4$$

$$y = \frac{-4}{2} + \frac{3}{2} = -\frac{1}{2}$$

$$\underline{x = -4, y = -\frac{1}{2}}$$

503. $5y + 9x = 53$

$$15y - 5x = 15$$

$$\frac{15y}{15} - \frac{5x}{15} = \frac{15}{15}$$

$$y - \frac{x}{3} = 1$$

$$y = \frac{x}{3} + 1$$

$$5\left(\frac{x}{3} + 1\right) + 9x = 53$$

$$\frac{5x}{3} + 5 + 9x = 53$$

$$\frac{32x}{3} = 43 - 5$$

$$\frac{32x}{3} = 48$$

$$\frac{\frac{32x}{3}}{\frac{32}{3}} = \frac{48}{\frac{32}{3}}$$

$$x = 4\frac{1}{2} = \frac{9}{2}$$

$$y = \frac{\frac{9}{2}}{3} + 1 = 2\frac{1}{2}$$

$$x = 4\frac{1}{2}, y = 2\frac{1}{2}$$

504. $25y + 12x = 3$

$$20y + 10x = 10$$

$$\frac{20y}{20} + \frac{10x}{20} = \frac{10}{20}$$

$$y + \frac{x}{2} = \frac{1}{2}$$

$$y = -\frac{x}{2} + \frac{1}{2}$$

$$25\left(-\frac{x}{2} + \frac{1}{2}\right) + 12x = 3$$

$$-\frac{25x}{2} + \frac{25}{2} + 12x = 3$$

$$-\frac{25x}{2} + \frac{24x}{2} = \frac{6}{2} - \frac{25}{2}$$

$$-\frac{x}{2} = -\frac{19}{2}$$

$$\frac{-\frac{x}{2}}{-\frac{1}{2}} = \frac{-\frac{19}{2}}{-\frac{1}{2}}$$

$$x = 19$$

$$y = -\frac{19}{2} + \frac{1}{2} = -9$$

$$x = 19, y = -9$$

505. $6y - 12x = 8$

$$8y - 12x = 20$$

$$\frac{6y}{6} - \frac{12x}{6} = \frac{8}{6}$$

$$y - 2x = \frac{4}{3}$$

$$y = 2x + \frac{4}{3}$$

$$8\left(2x + \frac{4}{3}\right) - 12x = 20$$

$$16x + \frac{32}{3} - 12x = 20$$

$$4x = \frac{60}{3} - \frac{32}{3}$$

$$4x = \frac{28}{3}$$

$$\frac{4x}{4} = \frac{\frac{28}{3}}{4}$$

$$x = \frac{7}{3}$$

$$y = 2\left(\frac{7}{3}\right) + \frac{4}{3}$$

$$y = \frac{14}{3} + \frac{4}{3} = \frac{18}{3} = 6$$

$$\underline{x = \frac{7}{3}, y = 6}$$

506. $20y + 5x = 10$

$$20y + 9x = 8$$

$$\frac{20y}{20} + \frac{5x}{20} = \frac{10}{20}$$

$$y + \frac{x}{4} = \frac{1}{2}$$

$$y = \frac{1}{2} - \frac{x}{4}$$

$$20\left(\frac{1}{2} - \frac{x}{4}\right) + 9x = 8$$

$$10 - 5x + 9x = 8$$

$$4x = 8 - 10$$

$$4x = -2$$

$$-6x = -6$$

$$-\frac{10x}{2} + \frac{10}{2} - 6x = -6$$

$$-5x + 5 - 6x = -6$$

$$-11x = -6 - 5$$

$$-11x = -11$$

$$\frac{-11x}{-11} = \frac{-11}{-11}$$

$$\frac{4x}{4} = \frac{-2}{4}$$

$$x = -\frac{1}{2}$$

$$y = \frac{1}{2} - \frac{-\frac{1}{2}}{4} = \frac{4}{8} + \frac{1}{8} = \frac{5}{8}$$

$$\underline{x = -\frac{1}{2}, y = \frac{5}{8}}$$

507. $12y + 10x = 56$

$$14y - 14x = 14$$

$$\frac{14y}{14} - \frac{14x}{14} = \frac{14}{14}$$

$$y - x = 1$$

$$y = x + 1$$

$$12(x + 1) + 10x = 56$$

$$12x + 12 + 10x = 56$$

$$22x = 56 - 12$$

$$22x = 44$$

$$\frac{22x}{22} = \frac{44}{22}$$

$$x = 2$$

$$y = 2 + 1 = 3$$

$$\underline{x = 2, y = 3}$$

508. $8y + 4x = 4$

$$10y - 6x = -6$$

$$\frac{8y}{8} + \frac{4x}{8} = \frac{4}{8}$$

$$y + \frac{x}{2} = \frac{1}{2}$$

$$y = -\frac{x}{2} + \frac{1}{2}$$

$$10\left(-\frac{x}{2} + \frac{1}{2}\right)$$

$$x = 1$$

$$y = -\frac{1}{2} + \frac{1}{2} = 0$$

$$\underline{x = 1, y = 0}$$

509. $6y + 9x = 2$

$$4y - 8x = 6$$

$$\frac{4y}{4} - \frac{8x}{4} = \frac{6}{4}$$

$$y - 2x = \frac{3}{2}$$

$$y = 2x + \frac{3}{2}$$

$$6\left(2x + \frac{3}{2}\right) + 9x = 2$$

$$12x + 9 + 9x = 2$$

$$21x = 2 - 9$$

$$21x = -7$$

$$\frac{21x}{21} = \frac{-7}{21}$$

$$x = -\frac{1}{3}$$

$$y = 2\left(-\frac{1}{3}\right) + \frac{3}{2} = -\frac{2}{3} + \frac{3}{2}$$

$$= -\frac{4}{6} + \frac{9}{6} = \frac{5}{6}$$

$$\underline{x = -\frac{1}{3}, y = \frac{5}{6}}$$

510. $10y + 5x = -10$

$$16y - 7x = 14$$

$$\frac{10y}{10} + \frac{5x}{10} = \frac{-10}{10}$$

$$y + \frac{x}{2} = -1$$

$$y = -\frac{x}{2} - 1$$

$$16\left(-\frac{x}{2} - 1\right) - 7x = 14$$

$$-8x - 16 - 7x = 14$$

$$-15x = 14 + 16$$

$$-15x = 30$$

$$\frac{-15x}{-15} = \frac{30}{-15}$$

$$x = -2$$

$$y = -\frac{-2}{2} - 1 = 0$$

$$\underline{x = -2, y = 0}$$

Solve each system of linear equations using addition method:

511. $2y + 2x = 4$
$-2y + x = 2$

$2y + 2x = 4$

$\underline{-2y + x = 2}$
$\qquad\quad 3x = 6$
$\qquad\quad \dfrac{3x}{3} = \dfrac{6}{3}$
$x = 2$
$2y + 2(2) = 4$
$2y + 4 = 4$
$2y = 4 - 4$
$y = 0$
$\underline{x = 2, y = 0}$

512. $-3y - 3x = 9$
$3y + 2x = 1$

$-3y - 3x = 9$
$\underline{+3y + 2x = 1}$
$-x = 10$
$\dfrac{-x}{-1} = \dfrac{10}{-1}$
$x = -10$
$-3y - 3(-10) = 9$
$-3y + 30 = 9$
$-3y = -21$
$\dfrac{-3y}{-3} = \dfrac{-21}{-3}$
$y = 7$
$\underline{x = -10, y = 7}$

513. $4y + 2x = 8$
$-4y - x = 1$

$4y + 2x = 8$

$\underline{-4y - x = 1}$
$x = 9$
$-4y - (9) = 1$
$-4y = 9 + 1$
$\dfrac{-4y}{-4} = \dfrac{10}{-4}$
$y = -\dfrac{10}{4} = -\dfrac{5}{2}$
$\underline{x = 9, y = -\dfrac{5}{2}}$

514. $3y - 7x = 5$
$-3y + x = 7$

$3y - 7x = 5$
$\underline{-3y + x = 7}$
$-6x = 12$
$\dfrac{-6x}{-6} = \dfrac{12}{-6}$
$x = -2$
$-3y - 2 = 7$
$-3y = 9$
$\dfrac{-3y}{-3} = \dfrac{9}{-3}$
$y = -3$
$x = -2, y = -3$

515. $-6y - 3x = -9$
$6y - x = 5$

$-6y - 3x = -9$
$\underline{6y - x = 5}$
$-4x = -4$
$\dfrac{-4x}{-4} = \dfrac{-4}{-4}$

$x = 1$

$6y - (1) = 5$

$6y = 5 + 1$

$6y = 6$

$\dfrac{6y}{6} = \dfrac{6}{6}$

$y = 1$

$\underline{x = 1, y = 1}$

516. $7y + 5x = 17$

$-7y + 3x = 7$

$7y + 5x = 17$

$\underline{-7y + 3x = 7}$

$8x = 24$

$\dfrac{8x}{8} = \dfrac{24}{8}$

$x = 3$

$7y + 5(3) = 17$

$7y + 15 = 17$

$7y = 17 - 15$

$7y = 2$

$\dfrac{7y}{7} = \dfrac{2}{7}$

$y = \dfrac{2}{7}$

$\underline{x = 3, y = \dfrac{2}{7}}$

517. $6y + 3x = 3$

$-6y - 2x = 3$

$6y + 3x = 3$

$\underline{-6y - 2x = 3}$

$x = 6$

$6y + 3(6) = 3$

$6y + 18 = 3$

$6y = 3 - 18 = -15$

$\dfrac{6y}{6} = \dfrac{-15}{6}$

$y = -2\dfrac{1}{2}$

$\underline{x = 6, y = -2\dfrac{1}{2}}$

518. $8y - 2x = -3$

$-8y - 4x = 15$

$8y - 2x = -3$

$\underline{-8y - 4x = 15}$

$-6x = 12$

$\dfrac{-6x}{-6} = \dfrac{12}{-6}$

$x = -2$

$8y - 2(-2) = -3$

$8y + 4 = -3$

$8y = -3 - 4$

$8y = -7$

$\dfrac{8y}{8} = \dfrac{-7}{8}$

$y = -\dfrac{7}{8}$

$\underline{x = -2, y = -\dfrac{7}{8}}$

519. $9y + 5x = 14$

$-9y + 5x = 6$

$9y + 5x = 14$

$\underline{-9y + 5x = 6}$

$10x = 20$

$\dfrac{10x}{10} = \dfrac{20}{10}$

$x = 2$

$9y + 5(2) = 14$

$9y + 10 = 14$

$9y = 14 - 10$

$9y = 4$

$$\frac{9y}{9} = \frac{4}{9}$$

$$y = \frac{4}{9}$$

$$\underline{x = 2, y = \frac{4}{9}}$$

520. $-7y - 2x = 24$
$7y + 6x = -8$

$$-7y - 2x = 24$$
$$\underline{7y + 6x = -8}$$
$$4x = 16$$
$$\frac{4x}{4} = \frac{16}{4}$$
$$x = 4$$
$$-7y - 2(4) = 24$$
$$-7y - 8 = 24$$
$$-7y = 24 + 8 = 32$$
$$\frac{-7y}{-7} = \frac{32}{-7}$$
$$y = -\frac{32}{7}$$

$$\underline{x = 4, y = -\frac{32}{7}}$$

521. $4y - 3x = 1$
$-4y + 2x = 6$

$$4y - 3x = 1$$
$$\underline{-4y + 2x = 6}$$
$$-x = 7$$
$$\frac{-x}{-1} = \frac{7}{-1}$$
$$x = -7$$
$$4y - 3(-7) = 1$$
$$4y = 1 - 21 = -20$$
$$\frac{4y}{4} = \frac{-20}{4}$$
$$y = -5$$
$$\underline{x = -7, y = -5}$$

522. $-7y + 8x = 1$
$7y - 6x = 7$

$$-7y + 8x = 1$$
$$\underline{7y - 6x = 7}$$
$$2x = 8$$
$$\frac{2x}{2} = \frac{8}{2}$$
$$x = 4$$
$$-7y + 8(4) = 1$$
$$-7y + 32 = 1$$
$$-7y = 1 - 32 = -31$$
$$\frac{-7y}{-7} = \frac{-31}{-7}$$
$$y = \frac{31}{7}$$

$$\underline{x = 4, y = \frac{31}{7}}$$

523. $-6y + 4x = -5$
$6y + 7x = 16$

$$-6y + 4x = -5$$
$$\underline{6y + 7x = 16}$$
$$11x = 11$$
$$\frac{11x}{11} = \frac{11}{11}$$
$$x = 1$$
$$-6y + 4(1) = -5$$
$$-6y + 4 = -5$$
$$-6y = -5 - 4 = -9$$
$$\frac{-6y}{-6} = \frac{-9}{-6}$$
$$y = \frac{3}{2}$$

$$\underline{x = 1, y = \frac{3}{2}}$$

524. $9y - 3x = -42$

$-9y - 7x = 2$

$9y - 3x = -42$
$\underline{-9y - 7x = 2}$
$-10x = -40$
$\dfrac{-10x}{-10} = \dfrac{-40}{-10}$
$x = 4$
$9y - 3(4) = -42$
$9y = -42 + 12 = -30$
$\dfrac{9y}{9} = \dfrac{-30}{9}$
$y = -\dfrac{10}{3}$
$\underline{x = 4, y = -\dfrac{10}{3}}$

525. $-y + 9x = 16$
$\quad\quad y - x = 8$

$-y + 9x = 16$
$\underline{\quad\;\; y - x = 8}$
$8x = 24$
$\dfrac{8x}{8} = \dfrac{24}{8}$
$x = 3$
$-y + 9(3) = 16$
$-y + 27 = 16$
$-y = 16 - 27 = -11$
$y = 11$
$\underline{x = 3, y = 11}$

526. $2y - 2x = 7$
$\quad\quad -2y + 4x = 3$

$2y - 2x = 7$
$\underline{-2y + 4x = 3}$
$2x = 10$
$\dfrac{2x}{2} = \dfrac{10}{2}$

$x = 5$
$2y - 2(5) = 7$
$2y = 7 + 10 = 17$
$\dfrac{2y}{2} = \dfrac{17}{2}$
$y = \dfrac{17}{2}$
$\underline{x = 5, y = \dfrac{17}{2}}$

527. $8y + 9x = 17$
$\quad\quad -8y - 5x = -5$

$8y + 9x = 17$
$\underline{-8y - 5x = -5}$
$4x = 12$
$\dfrac{4x}{4} = \dfrac{12}{4}$
$x = 3$
$8y + 9(3) = 17$
$8y + 27 = 17$
$8y = 17 - 27 = -10$
$8y = -10$
$\dfrac{8y}{8} = -\dfrac{10}{8}$
$y = -\dfrac{5}{4}$
$\underline{x = 3, y = -\dfrac{5}{4}}$

528. $5y + 5x = 5$
$\quad\quad -5y - 6x = 6$

$5y + 5x = 5$
$\underline{-5y - 6x = 6}$
$-x = 11$
$\dfrac{-x}{-1} = \dfrac{11}{-1}$
$x = -11$
$5y + 5(-11) = 5$

$5y - 55 = 5$

$5y = 5 + 55 = 60$

$\dfrac{5y}{5} = \dfrac{60}{5}$

$y = 12$

$\underline{x = -11, y = 12}$

529. $7y - 6x = -4$

$-7y + 2x = 4$

$7y - 6x = -4$

$\underline{-7y + 2x = 4}$

$-4x = 0$

$\dfrac{-4x}{-4} = \dfrac{0}{-4}$

$x = 0$

$7y - 6(0) = -4$

$7y = -4$

$\dfrac{7y}{7} = -\dfrac{4}{7}$

$y = -\dfrac{4}{7}$

$\underline{x = 0, y = -\dfrac{4}{7}}$

530. $2y + 4x = 8$

$2y - 4x = 4$

$2y + 4x = 8$

$\underline{2y - 4x = 4}$

$4y = 12$

$\dfrac{4y}{4} = \dfrac{12}{4}$

$y = 3$

$2(3) + 4x = 8$

$6 + 4x = 8$

$4x = 8 - 6 = 2$

$\dfrac{4x}{4} = \dfrac{2}{4}$

$x = \dfrac{2}{4} = \dfrac{1}{2}$

$\underline{x = \dfrac{1}{2}, y = 3}$

531. $2y + 3x = 6$

$3y - 3x = 9$

$2y + 3x = 6$

$\underline{3y - 3x = 9}$

$5y = 15$

$\dfrac{5y}{5} = \dfrac{15}{5}$

$y = 3$

$2(3) + 3x = 6$

$6 + 3x = 6$

$3x = 6 - 6$

$3x = 0$

$\dfrac{3x}{3} = \dfrac{0}{3}$

$x = 0$

$\underline{x = 0, y = 3}$

532. $2y - 4x = 16$

$4y + 4x = 8$

$2y - 4x = 16$

$\underline{4y + 4x = 8}$

$6y = 24$

$\dfrac{6y}{6} = \dfrac{24}{6}$

$y = 4$

$2(4) - 4x = 16$

$8 - 4x = 16$

$-4x = 16 - 8 = 8$

$-4x = 8$

$\dfrac{-4x}{-4} = \dfrac{8}{-4}$

$x = -2$

$\underline{x = -2, y = 4}$

533. $4y - 3x = 5$
$3y + 3x = 9$

$4y - 3x = 5$
$\underline{3y + 3x = 9}$
$7y = 14$
$\dfrac{7y}{7} = \dfrac{14}{7}$
$y = 2$
$4(2) - 3x = 5$
$8 - 3x = 5$
$-3x = 5 - 8$
$-3x = -3$
$\dfrac{-3x}{-3} = \dfrac{-3}{-3}$
$x = 1$
$\underline{x = 1, y = 2}$

534. $2y - 2x = 20$
$8y + 2x = 10$

$2y - 2x = 20$
$\underline{8y + 2x = 10}$
$10y = 30$
$\dfrac{10y}{10} = \dfrac{30}{10}$
$y = 3$
$2(3) - 2x = 20$
$6 - 2x = 20$
$-2x = 20 - 6 = 14$
$\dfrac{-2x}{-2} = \dfrac{14}{-2}$
$x = -7$
$\underline{x = -7, y = 3}$

535. $-2y - 5x = 10$
$3y + 5x = -4$

$-2y - 5x = 10$
$\underline{3y + 5x = -4}$
$y = 6$
$-2(6) - 5x = 10$
$-12 - 5x = 10$
$-5x = 10 + 12 = 22$
$\dfrac{-5x}{-5} = \dfrac{22}{-5}$
$x = -\dfrac{22}{5}$
$\underline{x = -\dfrac{22}{5}, y = 6}$

536. $4y - 4x = -20$
$-2y + 4x = 8$

$4y - 4x = -20$
$\underline{-2y + 4x = 8}$
$2y = -12$
$\dfrac{2y}{2} = \dfrac{-12}{2}$
$y = -6$
$4(-6) - 4x = -20$
$-24 - 4x = -20$
$-4x = -20 + 24 = 4$
$\dfrac{-4x}{-4} = \dfrac{4}{-4}$
$x = -1$
$\underline{x = -1, y = -6}$

537. $-4y - 8x = 12$
$2y + 8x = -10$

$-4y - 8x = 12$
$\underline{2y + 8x = -10}$
$-2y = 2$
$\dfrac{-2y}{-2} = \dfrac{2}{-2}$
$y = -1$

$-4(-1) - 8x = 12$

$4 - 8x = 12$

$-8x = 12 - 4 = 8$

$\dfrac{-8x}{-8} = \dfrac{8}{-8}$

$x = -1$

$\underline{x = -1, y = -1}$

538. $6y + 6x = 18$

$-4y - 6x = -8$

$6y + 6x = 18$

$\underline{-4y - 6x = -8}$

$2y = 10$

$\dfrac{2y}{2} = \dfrac{10}{2}$

$y = 5$

$6(5) + 6x = 18$

$30 + 6x = 18$

$6x = 18 - 30 = -12$

$\dfrac{6x}{6} = \dfrac{-12}{6}$

$x = -2$

$\underline{x = -2, y = 5}$

539. $-5y - 5x = 25$

$10y + 5x = -15$

$-5y - 5x = 25$

$\underline{10y + 5x = -15}$

$5y = 10$

$\dfrac{5y}{5} = \dfrac{10}{5}$

$y = 2$

$-5(2) - 5x = 25$

$-10 - 5x = 25$

$-5x = 25 + 10 = 35$

$\dfrac{-5x}{-5} = \dfrac{35}{-5}$

$x = -7$

$\underline{x = -7, y = 2}$

540. $7y - 7x = 21$

$-6y + 7x = -24$

$7y - 7x = 21$

$\underline{-6y + 7x = -24}$

$y = -3$

$7(-3) - 7x = 21$

$-21 - 7x = 21$

$-7x = 21 + 21 = 42$

$\dfrac{-7x}{-7} = \dfrac{42}{-7}$

$x = -6$

$\underline{x = -6, y = -3}$

541. $8y + 6x = 24$

$-2y + 4x = 16$

$8y + 6x = 24$

$\underline{4(-2y + 4x = 16)}$

$8y + 6x = 24$

$\underline{-8y + 16x = 64)}$

$22x = 88$

$\dfrac{22x}{22} = \dfrac{88}{22}$

$x = 4$

$8y + 6(4) = 24$

$8y + 24 = 24$

$8y = 24 - 24 = 0$

$\dfrac{8y}{8} = \dfrac{0}{8}$

$y = 0$

$\underline{x = 4, y = 0}$

542. $-7y - 3x = 54$

$3y + 6x = 46$

$2(-7y - 3x = 54)$

$\underline{3y + 6x = 46}$

$8y - 3x = -44$

$-14y - 6x = 108$
$\underline{3y + 6x = 46}$
$-11y = 154$
$\dfrac{-11y}{-11} = \dfrac{154}{-11}$
$y = -14$
$-7(-14) - 3x = 54$
$98 - 3x = 54$
$-3x = 54 - 98 = -44$
$-3x = -44$
$\dfrac{-3x}{-3} = \dfrac{-44}{-3}$
$x = 14\dfrac{2}{3}$
$\underline{x = 14\dfrac{2}{3}, y = -14}$

$3y - 9x = -6$
$\underline{-3(8y - 3x = -44)}$

$3y - 9x = -6$
$\underline{-24y + 9x = 132)}$
$-21y = 126$
$\dfrac{-21y}{-21} = \dfrac{126}{-21}$
$y = -6$
$3(-6) - 9x = -6$
$-18 - 9x = -6$
$-9x = -6 + 18 = 12$
$\dfrac{-9x}{-9} = \dfrac{12}{-9}$
$x = -1\dfrac{1}{3}$
$\underline{x = -1\dfrac{1}{3}, y = 6}$

543. $6y - 3x = -24$
$7y - 3x = 28$

$6y - 3x = -24$
$\underline{-1(7y - 3x = 28)}$

$6y - 3x = -24$
$\underline{-7y + 3x = -28)}$
$-y = -52$
$\dfrac{-y}{-1} = \dfrac{-52}{-1}$
$y = 52$
$6(52) - 3x = -24$
$312 - 3x = -24$
$-3x = -24 - 312 = -336$
$\dfrac{-3x}{-3} = \dfrac{-336}{-3}$
$x = 112$
$\underline{x = 112, y = 52}$

544. $3y - 9x = -6$

545. $-5y - 15x = 17$
$-3y + 5x = 13$

$-5y - 15x = 17$
$\underline{3(-3y + 5x = 13)}$

$-5y - 15x = 17$
$\underline{-9y + 15x = 39}$
$-14y = 56$
$\dfrac{-14y}{-14} = \dfrac{56}{-14}$
$y = -4$

$-5(-4) - 15x = 17$
$20 - 15x = 17$
$-15x = 17 - 20 = -3$
$-15x = -3$
$\dfrac{-15x}{-15} = \dfrac{-3}{-15}$

$$x = \frac{3}{15} = \frac{1}{5}$$

$$\underline{x = \frac{1}{5}, y = -4}$$

$$x = -\frac{1}{5}$$

$$\underline{x = -\frac{1}{5}, y = -1}$$

546. $15y + 10x = 5$
$\quad\ 2y - 2x = 4$

$15y + 10x = 5$
$\underline{5(2y - 2x = 4)}$

$15y + 10x = 5$
$\underline{10y - 10x = 20}$
$25y = 25$
$\dfrac{25y}{25} = \dfrac{25}{25}$
$y = 1$
$2(1) - 2x = 4$
$2 - 2x = 4$
$-2x = 4 - 2 = 2$
$\dfrac{-2x}{-2} = \dfrac{2}{-2}$
$x = -1$

547. $-3y - 5x = 4$
$\quad\ -4y - 10x = 6$

$-2(-3y - 5x = 4)$
$\underline{-4y - 10x = 6}$

$6y + 10x = -8$
$\underline{-4y - 10x = 6}$
$2y = -2$
$\dfrac{2y}{2} = \dfrac{-2}{2}$
$y = -1$
$-3(-1) - 5x = 4$
$3 - 5x = 4$
$-5x = 4 - 3 = 1$
$\dfrac{-5x}{-5} = \dfrac{1}{-5}$

548. $7y - 14x = 19$
$\quad\ -8y - 14x = 4$

$7y - 14x = 19$
$\underline{-1(-8y - 14x = 4)}$

$7y - 14x = 19$
$\underline{8y + 14x = -4)}$
$15y = 15$
$\dfrac{15y}{15} = \dfrac{15}{15}$
$y = 1$
$7(1) - 14x = 19$
$7 - 14x = 19$
$-14x = 19 - 7 = 12$
$\dfrac{-14x}{-14} = \dfrac{12}{-14}$
$x = -\dfrac{6}{7}$
$\underline{x = -\dfrac{6}{7}, y = 1}$

549. $4y + 8x = 0$
$\quad\ 2y - 4x = 8$

$4y + 8x = 0$
$\underline{2(2y - 4x = 8)}$

$4y + 8x = 0$
$\underline{4y - 8x = 16)}$
$8y = 16$
$\dfrac{8y}{8} = \dfrac{16}{8}$
$y = 2$
$4(2) + 8x = 0$
$8 + 8x = 0$
$8x = -8$

$$\frac{8x}{8} = \frac{-8}{8}$$
$$x = -1$$
$$\underline{x = -1, y = 2}$$

550. $-4y + 4x = 15$
$6y + 8x = 16$

$-2(-4y + 4x = 15)$
$\underline{6y + 8x = 16}$

$8y - 8x = -30$
$\underline{6y + 8x = 16}$
$14y = -14$
$\frac{14y}{14} = \frac{-14}{14}$
$y = -1$
$-4(-1) + 4x = 15$
$4 + 4x = 15$
$4x = 15 - 4$
$4x = 11$
$\frac{4x}{4} = \frac{11}{4}$
$x = 2\frac{3}{4}$
$\underline{x = 2\frac{3}{4}, y = -1}$

551. $12y + 3x = 9$
$3y + 6x = -3$

$-2(12y + 3x = 9)$
$\underline{3y + 6x = -3}$

$-24y - 6x = -18$
$\underline{3y + 6x = -3}$
$-21y = -21$
$\frac{-21y}{-21} = \frac{-21}{-21}$
$y = 1$
$12(1) + 3x = 9$

$12 + 3x = 9$
$3x = 9 - 12 = -3$
$\frac{3x}{3} = \frac{-3}{3}$
$x = -1$
$\underline{x = -1, y = 1}$

552. $6y + 18x = 9$
$4y + 6x = -3$

$6y + 18x = 9$
$\underline{-3(4y + 6x = -3)}$

$6y + 18x = 9$
$\underline{-12y - 18x = 9}$
$-6y = 18$
$\frac{-6y}{-6} = \frac{18}{-6}$
$y = -3$
$6(-3) + 18x = 9$
$-18 + 18x = 9$
$18x = 9 + 18 = 27$
$\frac{18x}{18} = \frac{27}{18}$
$x = \frac{3}{2}$
$\underline{x = \frac{3}{2}, y = -3}$

553. $-12y + 20x = -8$
$-3y + 6x = -2$

$-12y + 20x = -8$
$\underline{-4(-3y + 6x = -2)}$

$-12y + 20x = -8$
$\underline{12y - 24x = 8}$
$-4x = 0$
$\frac{-4x}{-4} = \frac{0}{-4}$
$x = 0$

$-12y + 20(0) = -8$

$-12y = -8$

$\dfrac{-12y}{-12} = \dfrac{-8}{-12}$

$y = \dfrac{2}{3}$

$\underline{x = 0, y = \dfrac{2}{3}}$

554. $6y - 8x = 12$

$4y - 8x = 16$

$6y - 8x = 12$

$\underline{-1(4y - 8x = 16)}$

$6y - 8x = 12$

$\underline{-4y + 8x = -16}$

$2y = -4$

$\dfrac{2y}{2} = \dfrac{-4}{2}$

$y = -2$

$6(-2) - 8x = 12$

$-12 - 8x = 12$

$-8x = 12 + 12 = 24$

$\dfrac{-8x}{-8} = \dfrac{24}{-8}$

$x = -3$

$\underline{x = -3, y = -2}$

555. $-8y + 9x = 9$

$4y + 8x = 8$

$-8y + 9x = 9$

$\underline{2(4y + 8x = 8)}$

$-8y + 9x = 9$

$\underline{8y + 16x = 16}$

$25x = 25$

$\dfrac{25x}{25} = \dfrac{25}{25}$

$x = 1$

$-8y + 9(1) = 9$

$8y = 9 - 9 = 0$

$\dfrac{8y}{8} = \dfrac{0}{8}$

$y = 0$

$\underline{x = 1, y = 0}$

556. $-3y - 6x = 18$

$-3y + 3x = 9$

$-3y - 6x = 18$

$\underline{-1(-3y + 3x = 9)}$

$-3y - 6x = 18$

$\underline{3y - 3x = -9}$

$-9x = 9$

$\dfrac{-9x}{-9} = \dfrac{9}{-9}$

$x = -1$

$-3y - 6(-1) = 18$

$-3y + 6 = 18$

$-3y = 18 - 6 = 12$

$\dfrac{-3y}{-3} = \dfrac{12}{-3}$

$y = -4$

$\underline{x = -1, y = -4}$

557. $6y + 18x = 18$

$4y + 2x = 12$

$6y + 18x = 18$

$\underline{-9(4y + 2x = 12)}$

$6y + 18x = 18$

$\underline{-36y - 18x = -108}$

$-30y = -90$

$\dfrac{-30y}{-30} = \dfrac{-90}{-30}$

$y = 3$

$6(3) + 18x = 18$

$18 + 18x = 18$

$18x = 18 - 18 = 0$

$x = 0$

$\underline{x = 0, y = 3}$　　　　　　$\underline{x = -1, -3}$

558. $4y - 8x = 56$
$16y + 12x = 4$

$-4(4y - 8x = 56)$
$16y + 12x = 4$

$-16y + 32x = -224$
$\underline{16y + 12x = 4}$
$44x = -220$
$\dfrac{44x}{44} = \dfrac{-220}{44}$
$x = -5$
$4y - 8(-5) = 56$
$4y + 40 = 56$
$4y = 56 - 40 = 16$
$\dfrac{4y}{4} = \dfrac{16}{4}$
$y = 4$
$\underline{x = -5, y = 4}$

559. $-10y - 9x = -21$
$5y - 5x = 20$

$-10y - 9x = -21$
$\underline{2(5y - 5x = 20)}$

$-10y - 9x = -21$
$\underline{10y - 10x = 40}$
$-19x = 19$
$\dfrac{-19x}{-19} = \dfrac{19}{-19}$
$x = -1$
$-10y - 9(1) = 21$
$-10y - 9 = 21$
$-10y = 21 + 9 = 30$
$-10y = 30$
$\dfrac{-10y}{-10} = \dfrac{30}{-10}$
$y = -3$

560. $8y + 16x = 20$
$-4y + 6x = 4$

$8y + 16x = 20$
$2(-4y + 6x = 4)$

$8y + 16x = 20$
$\underline{-8y + 12x = 8}$
$28x = 28$
$\dfrac{28x}{28} = \dfrac{28}{28}$
$x = 1$
$8y + 16(1) = 20$
$8y = 20 - 16 = 4$
$\dfrac{8y}{8} = \dfrac{4}{8}$
$y = \dfrac{1}{2}$
$\underline{x = 1, y = \dfrac{1}{2}}$

561. $5y - 10x = 29$
$5y + 3x = 3$

$5y - 10x = 29$
$-1(5y + 3x = 3)$

$5y - 10x = 29$
$\underline{-5y - 3x = -3}$
$-13x = 26$
$\dfrac{-13x}{-13} = \dfrac{26}{-13}$
$x = -2$
$5y + 3(-2) = 3$
$5y - 6 = 3$
$5y = 3 + 6 = 9$
$\dfrac{5y}{5} = \dfrac{9}{5}$
$y = \dfrac{9}{5}$
$\underline{x = -2, y = \dfrac{9}{5}}$

562. $-7y - 8x = 5$
$-14y - 14x = 8$

$-2(-7y - 8x = 5)$
$\underline{-14y - 14x = 8}$

$14y + 16x = -10$
$\underline{-14y - 14x = 8}$
$2x = -2$
$\dfrac{2x}{2} = \dfrac{-2}{2}$
$x = -1$
$-7y - 8(-1) = 5$
$-7y + 8 = 5$
$-7y = 5 - 8 = -3$
$\dfrac{-7y}{-7} = \dfrac{-3}{-7}$
$y = \dfrac{3}{7}$
$\underline{x = -1, y = \dfrac{3}{7}}$

563. $-6y + 4x = 3$
$8y + 40x = -38$

$-10(-6y + 4x = 3)$
$\underline{8y + 40x = -38}$

$60y - 40x = -30$
$\underline{8y + 40x = -38}$
$68y = -68$
$\dfrac{68y}{68} = \dfrac{-68}{68}$
$y = -1$
$-6(-1) + 4x = 3$
$6 + 4x = 3$
$4x = 3 - 6 = -3$
$\dfrac{4x}{4} = \dfrac{-3}{4}$
$x = -\dfrac{3}{4}$

$\underline{x = -\dfrac{3}{4}, y = -1}$

564. $-9y + 27x = 60$
$6y + 12x = 20$

$2(-9y + 27x = 60)$
$\underline{3(6y + 12x = 20)}$

$-18y + 54x = 120$
$\underline{18y + 36x = 60}$
$90x = 180$
$\dfrac{90x}{90} = \dfrac{180}{90}$
$x = 2$
$-9y + 27(2) = 60$
$-9y + 54 = 60$
$-9y = 60 - 54 = 6$
$\dfrac{-9y}{-9} = \dfrac{6}{-9}$
$y = -\dfrac{2}{3}$
$\underline{x = 2, y = -\dfrac{2}{3}}$

565. $9y + 36x = 81$
$5y + 6x = 10$

$9y + 36x = 81$
$\underline{-6(5y + 6x = 10)}$

$9y + 36x = 81$
$\underline{-30y - 36x = -60}$
$-21y = 21$
$\dfrac{-21y}{-21} = \dfrac{21}{-21}$
$y = -1$
$9(-1) + 36x = 81$
$-9 + 36x = 81$
$36x = 81 + 9 = 90$
$\dfrac{36x}{36} = \dfrac{90}{36}$

$$x = \frac{10}{4} = \frac{5}{2}$$

$$\underline{x = \frac{5}{2}, y = -1}$$

$$x = \frac{132}{120} = \frac{11}{10}$$

$$\underline{x = \frac{11}{10}, y = -6}$$

566. $-2y + 6x = 2$

$3y + 3x = 9$

$-2y + 6x = 2$
$\underline{-2(3y + 3x = 9)}$

$-2y + 6x = 2$
$\underline{-6y - 6x = -18}$
$-8y = -16$
$\dfrac{-8y}{-8} = \dfrac{-16}{-8}$
$y = 2$
$-2(2) + 6x = 2$

$-4 + 6x = 2$
$6x = 2 + 4 = 6$
$\dfrac{6x}{6} = \dfrac{6}{6}$
$x = 1$
$\underline{x = 1, y = 2}$

567. $18y + 120x = 24$

$3y + 60x = 48$

$18y + 120x = 24$
$\underline{-2(3y + 60x = 48)}$

$18y + 120x = 24$
$\underline{-6y - 120x = -96}$
$12y = -72$
$\dfrac{12y}{12} = \dfrac{-72}{12}$
$y = -6$
$18(-6) + 120x = 24$
$-108 + 120x = 24$
$120x = 132$
$\dfrac{120x}{120} = \dfrac{132}{120}$

568. $120y + 60x = 80$

$60y + 20x = -20$

$120y + 60x = 80$
$\underline{-2(60y + 20x = -20)}$

$120y + 60x = 80$
$\underline{-120y - 40x = 40}$
$20x = 120$
$\dfrac{20x}{20} = \dfrac{120}{20}$
$x = 6$
$120y + 60(6) = 80$
$120y + 360 = 80$
$120y = 80 - 360 = -280$
$\dfrac{120y}{120} = \dfrac{280}{120}$
$y = -\dfrac{280}{120} = -\dfrac{7}{3}$
$\underline{x = 6, y = -\dfrac{7}{3}}$

569. $-14y - 21x = -7$

$-5y - 7x = 20$

$-14y - 21x = -7$
$\underline{-3(-5y - 7x = 20)}$

$-14y - 21x = -7$
$\underline{15y + 21x = -60}$
$y = -67$
$-14(-67) - 21x = -7$
$938 - 21x = -7$
$-21x = -7 - 938$
$-21x = -945$
$\dfrac{-21x}{-21} = \dfrac{-945}{-21}$

$x = 45$
$\underline{x = 45, y = -67}$

570. $-8y + 13x = 17$
$\quad 8y - 8x = 8$

$-8y + 13x = 17$
$\underline{8y - 8x = 8}$
$5x = 25$
$\dfrac{5x}{5} = \dfrac{25}{5}$
$x = 5$
$8y - 8(5) = 8$
$8y - 40 = 8$
$8y = 8 + 40 = 48$
$\dfrac{8y}{8} = \dfrac{48}{8}$
$y = 6$
$\underline{x = 5, y = 6}$

571. $4y + 6x = 30$
$\quad 6y - 2x = 12$

$4y + 6x = 30$
$\underline{3(6y - 2x = 12)}$

$4y + 6x = 30$
$\underline{18y - 6x = 36}$
$22y = 66$
$\dfrac{22y}{22} = \dfrac{66}{22}$
$y = 3$
$4(3) + 6x = 30$
$12 + 6x = 30$
$6x = 30 - 12 = 18$
$\dfrac{6x}{6} = \dfrac{18}{6}$
$x = 3$
$\underline{x = 3, y = 3}$

572. $-9y - 4x = 8$
$\quad -12y - 8x = 10$

$-2(-9y - 4x = 8)$
$\underline{-12y - 8x = 10}$

$18y + 8x = -16$
$\underline{-12y - 8x = 10}$
$6y = -6$
$\dfrac{6y}{6} = \dfrac{-6}{6}$
$y = -1$
$-9(-1) - 4x = 8$
$9 - 4x = 8$
$-4x = 8 - 9 = -1$
$\dfrac{-4x}{-4} = \dfrac{-1}{-4}$
$x = \dfrac{1}{4}$
$\underline{x = \dfrac{1}{4}, y = -1}$

573. $10y + 6x = 17$
$\quad 5y - 4x = 12$

$2(10y + 6x = 17)$
$\underline{3(5y - 4x = 12)}$

$20y + 12x = 34$
$\underline{15y - 12x = 36}$
$35y = 70$
$\dfrac{35y}{35} = \dfrac{70}{35}$
$y = 2$
$10(2) + 6x = 17$
$20 + 6x = 17$
$6x = 17 - 20 = -3$
$\dfrac{6x}{6} = -\dfrac{3}{6}$
$x = -\dfrac{1}{2}$
$\underline{x = 1\dfrac{1}{2}, y = 2}$

574. $23y + 12x = 27$
$\quad 16y - 36x = 4$

$3(23y + 12x = 27)$

$\underline{16y - 36x = 4}$

$69y + 36x = 81$

$\underline{16y - 36x = 4}$

$85y = 85$

$\dfrac{85y}{85} = \dfrac{85}{85}$

$y = 1$

$16(1) - 36x = 4$

$-36x = 4 - 16 = -12$

$\dfrac{-36x}{-36} = \dfrac{-12}{-36}$

$x = \dfrac{1}{3}$

$x = \dfrac{1}{3}, y = 1$

575. $9y + 18x = 23$

$-16y + 9x = -9$

$9y + 18x = 23$

$\underline{-2(-16y + 9x = -9)}$

$9y + 18x = 23$

$\underline{32y - 18x = 18}$

$41y = 41$

$\dfrac{41y}{41} = \dfrac{41}{41}$

$y = 1$

$9(1) + 18x = 23$

$9 + 18x = 23$

$18x = 23 - 9 = 14$

$\dfrac{18x}{18} = \dfrac{14}{18}$

$x = \dfrac{7}{9}$

$x = \dfrac{7}{9}, y = 1$

576. $5y - 15x = 31$

$7y - 5x = 5$

$5y - 15x = 31$

$\underline{-3(7y - 5x = 5)}$

$5y - 15x = 31$

$\underline{-21y + 15x = -15}$

$-16y = 16$

$\dfrac{-16y}{-16} = \dfrac{16}{-16}$

$y = -1$

$5(-1) - 15x = 31$

$-5 - 15x = 31$

$-15x = 31 + 5 = 36$

$\dfrac{-15x}{-15} = \dfrac{36}{-15}$

$x = -\dfrac{12}{5}$

$x = -\dfrac{12}{5}, y = -1$

577. $-6y + 7x = 10$

$9y + 6x = 18$

$9(-6y + 7x = 10)$

$\underline{+6(9y + 6x = 18)}$

$-54y + 63x = 90$

$\underline{+54y + 36x = 108}$

$99x = 198$

$\dfrac{99x}{99} = \dfrac{198}{99}$

$x = 2$

$-6y + 7(2) = 10$

$-6y + 14 = 10$

$-6y = 10 - 14 = -4$

$\dfrac{-6y}{-6} = \dfrac{-4}{-6}$

$y = \dfrac{4}{6} = \dfrac{2}{3}$

$$x = 2, y = \frac{2}{3}$$

$$y = 6$$
$$x = 1, y = 6$$

578. $-2y + 12x = 16$
$8y - 9x = 14$

$4(-2y + 12x = 16)$
$\underline{+8y - 9x = 14}$

$-8y + 48x = 64$
$\underline{+8y - 9x = 14}$
$39x = 78$
$\dfrac{39x}{39} = \dfrac{78}{39}$
$x = 2$
$-2y + 12(2) = 16$
$-2y + 24 = 16$
$-2y = 16 - 24 = -8$
$\dfrac{-2y}{-2} = \dfrac{-8}{-2}$
$y = 4$
$x = 2, y = 4$

579. $5y - 25x = 5$
$3y + 15x = 33$

$-3(5y - 25x = 5)$
$\underline{+5(3y + 15x = 33)}$

$-15y + 75x = -15$
$\underline{+15y + 75x = 165}$
$150x = 150$
$\dfrac{150x}{150} = \dfrac{150}{150}$
$x = 1$
$5y - 25(1) = 5$
$5y - 25 = 5$
$5y = 5 + 25 = 30$
$\dfrac{5y}{5} = \dfrac{30}{5}$

580. $2y + 12x = 12$
$2y + 9x = 9$

$2y + 12x = 12$
$-1(2y + 9x = 9)$

$2y + 12x = 12$
$\underline{-2y - 9x = -9}$
$3x = 3$
$\dfrac{3x}{3} = \dfrac{3}{3}$
$x = 1$
$2y + 12(1) = 12$
$2y = 12 - 12 = 0$
$\dfrac{2y}{2} = \dfrac{0}{2}$
$y = 0$

581. $9y + 8x = 8$
$\underline{18y + 2x = 30}$

$-2(9y + 8x = 8)$
$\underline{18y + 2x = 30}$

$-18y - 16x = -16$
$\underline{18y + 2x = 30}$
$-14x = 14$
$\dfrac{-14x}{-14} = \dfrac{14}{-14}$
$x = -1$
$9y + 8(-1) = 8$
$9y - 8 = 8$
$9y = 8 + 8 = 16$
$\dfrac{9y}{9} = \dfrac{16}{9}$
$y = \dfrac{16}{9}$
$x = -1, y = \dfrac{16}{9}$

582. $15y - 9x = 6$
$10y - 5x = 15$

$-2(15y - 9x = 6)$
$+3(10y - 5x = 15)$

$-30y + 18x = -12$
$+30y - 15x = 45$
$3x = 33$
$\dfrac{3x}{3} = \dfrac{33}{3}$
$x = 11$
$15y - 9(11) = 6$
$15y - 99 = 6$
$15y = 6 + 99 = 105$
$\dfrac{15y}{15} = \dfrac{105}{15}$
$y = 7$
$\underline{x = 11, y = 7}$

583. $-5y + 9x = 12$
$15y - 5x = 30$

$3(-5y + 9x = 12)$
$15y - 5x = 30$

$-15y + 27x = 36$
$15y - 5x = 30$
$22x = 66$
$\dfrac{22x}{22} = \dfrac{66}{22}$
$x = 3$
$-5y + 9(3) = 12$
$-5y + 27 = 12$
$-5y = 12 - 27 = -15$
$\dfrac{-5y}{-5} = \dfrac{-15}{-5}$
$y = 3$
$\underline{x = 3, y = 3}$

584. $-25y + 20x = 10$

$20y + 4x = 12$

$4(-25y + 20x = 10)$
$+5(20y + 4x = 12)$

$-100y + 80x = 40$
$+100y + 20x = 60$
$100x = 100$
$\dfrac{100x}{100} = \dfrac{100}{100}$
$x = 1$
$-25y + 20(1) = 10$
$-25y + 20 = 10$
$-25y = 10 - 20 = -10$
$\dfrac{-25y}{-25} = \dfrac{-10}{-25}$
$y = \dfrac{2}{5}$
$\underline{x = 1, y = \dfrac{2}{5}}$

585. $6y - 12x = 8$
$3y - 12x = 20$

$6y - 12x = 8$
$-1(3y - 12x = 20)$

$6y - 12x = 8$
$-3y + 12x = -20$
$3y = -12$
$\dfrac{3y}{3} = \dfrac{-12}{3}$
$y = -4$
$6(-4) - 12x = 8$
$-24 - 12x = 8$
$-12x = 8 + 24 = 32$
$\dfrac{-12x}{-12} = \dfrac{32}{-12}$
$x = -\dfrac{32}{12} = -\dfrac{8}{3}$
$\underline{x = -\dfrac{8}{3}, y = -4}$

586. $-20y + 5x = 10$
$20y + 9x = 4$

$-20y + 5x = 10$
$\underline{20y + 9x = 4}$
$14x = 14$
$\dfrac{14x}{14} = \dfrac{14}{14}$
$x = 1$
$-20y + 5(1) = 10$
$-20y + 5 = 10$
$-20y = 10 - 5 = 5$
$\dfrac{-20y}{-20} = \dfrac{5}{-20}$
$y = -\dfrac{1}{4}$
$\underline{x = 1, y = -\dfrac{1}{4}}$

587. $4y + 10x = 14$
$16y - 14x = 2$

$-4(4y + 10x = 14)$
$\underline{16y - 14x = 2}$

$-16y - 40x = -56$
$\underline{16y - 14x = 2}$
$\dfrac{-54x}{-54} = \dfrac{-54}{-54}$
$x = 1$
$4y + 10(1) = 14$
$4y + 10 = 14$
$4y = 14 - 10 = 4$
$\dfrac{4y}{4} = \dfrac{4}{4}$
$y = 1$
$\underline{x = 1, y = 1}$

588. $8y + 4x = 9$
$-10y - 8x = -6$

$2(8y + 4x = 9)$

$-10y - 8x = -6$

$16y + 8x = 18$
$\underline{-10y - 8x = -6}$
$6y = 12$
$\dfrac{6y}{6} = \dfrac{12}{6}$
$y = 2$
$8(2) + 4x = 9$
$16 + 4x = 9$
$4x = 9 - 16 = -7$
$\dfrac{4x}{4} = \dfrac{-7}{4}$
$x = -\dfrac{7}{4}$
$\underline{x = -\dfrac{7}{4}, y = 2}$

589. $6y + 9x = 2$
$-4y - 8x = 6$

$4(6y + 9x = 2)$
$\underline{6(-4y - 8x = 6)}$

$24y + 36x = 8$
$\underline{-24y - 48x = 36}$
$-12x = 44$
$\dfrac{-12x}{-12} = \dfrac{44}{-12}$
$x = -\dfrac{44}{12} = -\dfrac{11}{3}$
$6y + 9\left(-\dfrac{11}{3}\right) = 2$
$6y - \dfrac{99}{3} = 2$
$6y - 33 = 2$
$6y = 2 + 33 = 35$
$6y = 35$
$\dfrac{6y}{6} = \dfrac{35}{6}$
$y = \dfrac{35}{6}$
$\underline{x = -\dfrac{11}{3}, y = \dfrac{35}{6}}$

590. $10y + 5x = -10$
$16y - 7x = 14$

$-8(10y + 5x = -10)$
$\underline{5(16y - 7x = 14)}$

$-80y - 40x = 80$
$\underline{80y - 35x = 70}$
$-75x = 150$

$\dfrac{-75x}{-75} = \dfrac{150}{-75}$
$x = -2$
$10y + 5(-2) = -10$
$10y - 10 = -10$
$10y = -10 + 10 = 0$
$\dfrac{10y}{10} = \dfrac{0}{10}$
$y = 0$
$\underline{x = -2, y = 0}$

Answers to Chapter 7 Polynomials

Add the below polynomials:

591. $(3 + 3x) + (1 + 2x) = 4 + 5x$

592. $(4 + 2x) + (2 + 5x) = 6 + 7x$

593. $(7 + 2x) + (1 + 7x) = 8 + 9x$

594. $(1 + 9x) + (8 + 4x) = 9 + 13x$

595. $(9 + 2x) + (3 + 3x) = 12 + 5x$

596. $(12 + 8x) + (9 + 2x) = 21 + 10x$

597. $(5 + 12x) + (4 + 12x) = 9 + 24x$

598. $(8 + 7x) + (7 + 13x) = 15 + 20x$

599. $(11 + 6x) + (2 + 6x) = 13 + 12x$

600. $(2 + x) + (11 + 9x) = 13 + 10x$

601. $(6 + 5x) + (5 + 2x) = 11 + 7x$

602. $(1 + 9x) + (5 + 6x) = 6 + 15x$

603. $(8 + 6x) + (10 + 8x) = 18 + 14x$

604. $(7 + 2x) + (6 + 3x) = 13 + 5x$

605. $(3 + 6x + 3x^2) + (14 + 15x + x^2) = 17 + 21x + 4x^2$

606. $(12 + 4x + 10x^2) + (3 + 3x + 4x^2) = 15 + 7x + 14x^2$

607. $(1 + 5x + 12x^2) + (18 + 6x + 13x^2) = 19 + 11x + 25x^2$

608. $(6 + 12x + 12x^2) + (11 + 4x + 15x^2) = 17 + 16x + 27x^2$

609. $(10 + 6x + 9x^2) + (4 + 11x + 10x^2) = 14 + 17x + 19x^2$

610. $(3 + 11x + 4x^2) + (5 + x + 8x^2) = 8 + 12x + 12x^2$

611. $(9 + x + 6x^2) + (21 + 7x + 5x^2) = 30 + 8x + 11x^2$

612. $(5 + 18x + 10x^2) + (5 + 6x + 12x^2) = 10 + 24x + 22x^2$

613. $(7 + 17x + 7x^2) + (17 + 12x + 12x^2) = 24 + 29x + 19x^2$

614. $(8 + 7x + 9x^2) + (6 + 17x + 7x^2) = 14 + 24x + 16x^2$

615. $(9 + 13x + 8x^2) + (7 + 8x + 10x^2) = 16 + 21x + 18x^2$

616. $(12 + 15x - 8x^2) + (12 + 4x + 9x^2) = 24 + 19x + x^2$

617. $(6 - 24x + 3x^2) + (4 + 9x + 2x^2) = 10 - 15x + 5x^2$

618. $(5 + 4x - 12x^2) + (9 - 10x + 3x^2) = 14 - 6x - 9x^2$

619. $(11 + 9x + 6x^2) + (9 + 13x + 4x^2) = 20 + 22x + 10x^2$

620. $(15 - 5x - 11x^2) + (7 - 8x - 10x^2) = 22 - 13x - 21x^2$

621. $(4 + 8x - 10x^2) + (10 - 2x + 6x^2) = 14 + 6x - 4x^2$

622. $(3 + 6x + 7x^2) + (6 + 6x + 11x^2) = 9 + 12x + 18x^2$

623. $(4 + 3x - 4x^2) + (12 - 4x + x^2) = 16 - x - 3x^2$

624. $(2 - 12x + 5x^2) + (8 + 12x + 7x^2) = 10 + 12x^2$
625. $(18 + 20x - 4x^2) + (3 + 3x + 13x^2) = 21 + 23x + 9x^2$

Subtract the below polynomials:

626. $(3 + 6x + 3x^2) - (14 + 15x + x^2) = -11 - 9x + 2x^2$
627. $(12 + 4x + 10x^2) - (3 + 3x + 4x^2) = 9 + x + 6x^2$
628. $(1 + 5x + 12x^2) - (18 + 6x + 13x^2) = -17 - x - x^2$
629. $(6 + 12x + 12x^2) - (11 + 4x + 15x^2) = -5 + 8x - 3x^2$
630. $(10 + 6x + 9x^2) - (4 + 11x + 10x^2) = 6 - 5x - x^2$
631. $(3 + 11x + 4x^2) - (5 + x + 8x^2) = -2 + 10x - 4x^2$
632. $(9 + x + 6x^2) - (21 + 7x + 5x^2) = -12 - 6x + x^2$
633. $(5 + 18x + 10x^2) - (5 + 6x + 12x^2) = 12x - 2x^2$
634. $(7 + 17x + 7x^2) - (17 + 12x + 12x^2) = -10 + 5x - 5x^2$
635. $(8 + 7x + 9x^2) - (6 + 17x + 7x^2) = 2 - 10x + 2x^2$
636. $(9 + 13x + 8x^2) - (7 + 8x + 10x^2) = 2 + 5x - 2x^2$
637. $(12 + 15x - 8x^2) - (12 + 4x + 9x^2) = 11x - 17x^2$
638. $(6 - 24x + 3x^2) - (4 + 9x + 2x^2) = 2 - 33x + x^2$
639. $(5 + 4x - 12x^2) - (9 - 10x + 3x^2) = -4 + 14x - 15x^2$
640. $(11 + 9x + 6x^2) - (9 + 13x + 4x^2) = 2 - 4x + 2x^2$
641. $(15 - 5x - 11x^2) - (7 - 8x - 10x^2) = 8 + 3x - x^2$
642. $(4 + 8x - 10x^2) - (10 - 2x + 6x^2) = -6 + 10x - 16x^2$
643. $(3 + 6x + 7x^2) - (6 + 6x + 11x^2) = -3 - 4x^2$
644. $(4 + 3x - 3x^2) - (12 - 4x + x^2) = -8 + 7x - 4x^2$
645. $(2 - 12x + 5x^2) - (8 + 12x + 7x^2) = -6 - 24x - 2x^2$
646. $(18 + 20x - 4x^2) - (3 + 3x + 13x^2) = 15 + 17x - 17x^2$

Simplify the below expression:

647. $(2x^3 3yz^2)(2x^2 3y^3 z^3)(x^2 y^3 4z^5)$
$$(2 \times 3 \times 2 \times 3 \times 4x^{3+2+2} y^{1+3+3} z^{2+3+5})$$
$$144x^7 y^7 z^{10}$$

648. $(3x^2 y^2 2z^3)^3 (2x^5 y^4 z^5)(x^7 2y^3 2z^4)^2$
$$(3^3 x^{2\times3} y^{2\times3} 2^3 z^{3\times3})(2x^5 y^4 z^5)(x^{7\times2} 2^2 y^{3\times2} 2^2 z^{4\times2})$$
$$(27x^6 y^6 8z^9)(2x^5 y^4 z^5)(x^{14} 4y^6 4z^8)$$
$$(27 \times 8 \times 2 \times 4 \times 4x^{6+5+14} y^{6+4+6} z^{9+5+8})$$
$$(6{,}912x^{25} y^{16} z^{22})$$

649. $(x^44y^33z^4)(x^4y^5z^2)(3x^62y^8z^2)^3$

$$(x^44y^33z^4)(x^4y^5z^2)(3x^62y^8z^2)^3$$

$$(x^44y^33z^4)(x^4y^5z^2)(3^3x^{6\times3}2^3y^{8\times3}z^{2\times3})$$

$$(x^44y^33z^4)(x^4y^5z^2)(27x^{18}8y^{24}z^6)$$

$$(4\times3\times27\times8x^{4+4+18}y^{3+5+24}z^{4+2+6})$$

$$2{,}592x^{26}y^{33}z^{14}$$

650. $(x2y^44z^5)^2(x^43y^2z)^3(2x^2y^3z)^2$

$$(x2y^44z^5)^2(x^43y^2z)^3(2x^2y^3z)^2$$

$$(x^22^2y^{4\times2}4^2z^{5\times2})(x^{4\times3}3^3y^{2\times3}z^3)(2^2x^{2\times2}y^{3\times2}z^2)$$

$$(x^24y^816z^{10})(x^{12}27y^6z^3)(4x^4y^6z^2)$$

$$(4\times16\times27\times4x^{2+12+}\ y^{8+6+6}z^{10+3+2})$$

$$6{,}912x^{18}y^{20}z^{15}$$

651. $(x^4y^57z^6)(4x^7y^7z^4)(2x^2y^5z^2)^2$

$$(x^4y^57z^6)(4x^7y^7z^4)(2^2x^{2\times2}y^{5\times2}z^{2\times2})$$

$$(x^4y^57z^6)(4x^7y^7z^4)(4x^4y^{10}z^4)$$

$$(7\times4\times4x^{4+7+4}y^{5+7+10}z^{6+4+4})$$

$$112x^{15}y^{22}z^{14}$$

652. $(x^32y^{12}z^4)(x^92y^9z^6)(x^33y^83z^6)$

$$(2\times2\times3\times3x^{3+9+3}y^{12+9+8}z^{4+6+6})$$

$$(36x^{15}y^{29}z^{16})$$

653. $(x^4y^6z^7)^3(x^24y^4z^7)(x^25y^2z^4)^2$

$$(x^{4\times3}y^{6\times3}z^{7\times3})(x^24y^4z^7)(x^{2\times2}5^2y^{2\times2}z^{4\times2})$$

$$(x^{12}y^{18}z^{21})(x^24y^4z^7)(x^425y^4z^8)$$

$$(4\times25x^{12+2+4}y^{18+4+4}z^{21+7+8})$$

$$100x^{18}y^{26}z^{36}$$

654. $(xy^24z^2)(x^92y^3z^5)^4(x^55y^3z^1)$

$$(xy^24z^2)(x^{9\times4}2^4y^{3\times4}z^{5\times4})(x^55y^3z^1)$$

$$(xy^24z^2)(x^{36}16y^{12}z^{20})(x^55y^3z^1)$$

$$(4\times16\times5x^{1+36+}\ y^{2+12+3}z^{2+20+1})$$

$$320x^{42}y^{17}z^{23}$$

655. $(x^6y^32z^3)(x^52y^6z^2)(6x^62z^2)$

$$2 \times 2 \times 6 \times 2x^{6+5+6}y^{3+6}z^{3+2+2}$$

$$48x^{17}y^9z^7$$

656. $(x^25y^23z^6)(x^44y^7z)(x^4y^4z^4)^2$

$$(x^25y^23z^6)(x^44y^7z)(x^{4\times2}y^{4\times2}z^{4\times2})$$

$$(x^25y^23z^6)(x^44y^7z)(x^8y^8z^8)$$

$$5 \times 3 \times 4x^{2+4+8}y^{2+7+8}z^{6+1+8}$$

$$60x^{14}y^{17}z^{15}$$

657. $(4x^4y^75z^4)(x^62y^{11}z^5)(x^2y^2z^4)^2$

$$(4x^4y^75z^4)(x^64y^{11}z^5)(x^{2\times2}y^{2\times2}z^{4\times2})$$

$$(4x^4y^75z^4)(x^64y^{11}z^5)(x^4y^4z^8)$$

$$(4 \times 5 \times 4x^{4+6+4}y^{7+11+}\ z^{4+5+8})$$

$$80x^{14}y^{22}z^{17}$$

658. $(x^7y^4z^2)(x^7y^32z^6)^3(2x^3y^75z^8)^2$

$$(x^7y^4z^2)(x^{7\times3}y^{3\times3}2^3z^{6\times3})(2^2x^{3\times2}y^{7\times2}5^2z^{8\times2})$$

$$(x^7y^4z^2)(x^{21}y^98z^{18})(4x^6y^{14}25z^{16})$$

$$(8 \times 4 \times 25x^{7+21+6}y^{4+9+1}\ z^{2+18+16})$$

$$800x^{34}y^{27}z^{36}$$

659. $(2x^2y^5z^0)^4(x^24y^4z^8)(4x^{11}y^34z^6)$

$$(2^4x^{2\times4}y^{5\times4}z^{0\times4})(x^24y^4z^8)(4x^{11}y^34z^6)$$

$$(16x^8y^{20}z^0)(x^24y^4z^8)(4x^{11}y^34z^6)$$

$$(16 \times 4 \times 4 \times 4x^{8+2+1}\ y^{20+4+}\ z^{0+8+6})$$

$$1{,}024x^{21}y^{27}z^{14}$$

660. $(2x^42y3z^8)(x^46y^5z^7)(x^8y^32z^2)^2$

$$(2x^42y3z^8)(x^46y^5z^7)(x^{8\times2}y^{3\times2}2^2z^{2\times2})$$

$$(2x^42y3z^8)(x^46y^5z^7)(x^{16}y^64z^4)$$

$$(2 \times 2 \times 3 \times 6 \times 4x^{4+4+1}\ y^{1+5+6}z^{8+7+4})$$

$$288x^{24}y^{12}z^{19}$$

661. $(2x^5y^210z^4)(2x^3y^5z^2)(x^6y^43z^8)$

$$(2 \times 10 \times 2 \times 3x^{5+3+6}y^{2+5+4}z^{4+2+8})$$

$$120x^{14}y^{11}z^{14}$$

662. $(x^25y^7z^5)(8x^8z^4)(x^44y^6z^7)$

$$(5 \times 8 \times 4x^{2+8+4}y^{7+6}z^{5+4+7})$$

$$160x^{14}y^{13}z^{16}$$

663. $(10x^24y^3z^2)(x^2y^3z)(x^33y^9z^{12})$

$$(10 \times 4 \times 3x^{2+2+3}y^{3+3+9}z^{2+1+12})$$

$$120x^7y^{15}z^{15}$$

664. $(3x^2yz^5)^2(x^32y^2z^3)^2(x2y^{14}2z^8)^2$

$$(3^2x^{2\times2}y^2z^{5\times2})(x^{3\times2}2^2y^{2\times2}z^{3\times2})(x^22^2y^{14\times2}2^2z^{8\times2})$$

$$(9x^4y^2z^{10})(x^64y^4z^6)(x^24y^{28}4z^{16})$$

$$(9 \times 4 \times 4 \times 4x^{4+6+2}y^{2+4+2}\ z^{10+6+\ \ })$$

$$576x^{12}y^{34}z^{32}$$

665. $(x^46y^33z^8)(x^74yz^2)(x^7y^34z^4)$

$$(6 \times 3 \times 4x^{4+7}y^{3+1}z^{8+2})(x^7y^34z^4)$$

$$(72x^{11}y^4z^{10})(x^7y^34z^4)$$

$$(72 \times 4x^{11+7}y^{4+3}z^{10+4})$$

$$288x^{18}y^7z^{14}$$

666. $(x^5y^64z)(x^53y^3z)(2x^2y^23z^6)^3$

$$(x^5y^64z)(x^53y^3z)(2^3x^{2\times3}y^{2\times3}3^3z^{6\times3})$$

$$(x^5y^64z)(x^53y^3z)(8x^6y^627z^{18})$$

$$(4 \times 3 \times 8 \times 27x^{5+5+6}y^{6+3+6}z^{1+1+18})$$

$$2,592x^{16}y^{15}z^{20}$$

667. $(9x^2yz^4)(x^64y^6z^5)(x^4y^42z^8)^2$

$$(9x^2yz^4)(x^64y^6z^5)(x^{4\times2}y^{4\times2}2^2z^{8\times2})$$

$$(9x^2yz^4)(x^64y^6z^5)(x^8y^84z^{16})$$

$$(9 \times 4 \times 4x^{2+6+8}y^{1+6+8}z^{4+5+1}\ )$$

$$144x^{16}y^{15}z^{25}$$

668. $(x^4yz^2)(6x^2y^4z^3)(4x^3yz^2)$

$$(6 \times 4x^{4+2+3}y^{1+4+1}z^{2+3+2})$$

$$24x^9y^6z^7$$

669. $(2x^3y^5z^2)(x^67y^3z^5)^5(x^24y^{12}z^4)$

$$(2x^3y^5z^2)(x^{6\times5}7^5y^{3\times5}z^{5\times5})(x^24y^{12}z^4)$$

$$(2x^3y^5z^2)(x^{30}16{,}807y^{15}z^{25})(x^24y^{12}z^4)$$

$$(2\times16{,}807\times4x^{3+30+2}y^{5+15+12}z^{2+25+4})$$

$$134{,}456x^{35}y^{32}z^{31}$$

670. $(x4yz^2)(3x^6y^2z^6)(x^63y^4z^5)^4$

$$(x4yz^2)(3x^6y^2z^6)(x^{6\times4}3^4y^{4\times4}z^{5\times4})$$

$$(x4yz^2)(3x^6y^2z^6)(x^{24}81y^{16}z^{20})$$

$$(4\times3\times81x^{1+6+24}y^{1+2+16}z^{2+6+20})$$

$$972x^{31}y^{19}z^{28}$$

671. $(x^23y^5z^2)^4(x4y^8z^7)^3(3x^5y^5z^7)^2$

$$(x^{2\times4}3^4y^{5\times4}z^{2\times4})(x^34^3y^{8\times3}z^{7\times3})(3^2x^{5\times2}y^{5\times2}z^{7\times2})$$

$$(x^881y^{20}z^8)(x^364y^{24}z^{21})(9x^{10}y^{10}z^{14})$$

$$(81\times64\times9x^{8+3+10}y^{20+24+10}z^{8+21+14})$$

$$46{,}656x^{21}y^{54}z^{43}$$

672. $(x^22y^3z^2)^7(2x^3y^3z^8)^6(x^2y^22z^6)^3$

$$(x^{2\times7}2^7y^{3\times7}z^{2\times7})(2^6x^{3\times6}y^{3\times6}z^{8\times6})(x^{2\times3}y^{2\times3}2^3z^{6\times3})$$

$$(x^{14}128y^{21}z^{14})(64x^{18}y^{18}z^{48})(x^6y^68z^{18})$$

$$(128\times64\times8x^{14+18+6}y^{21+18}\quad z^{14+48+18})$$

$$(65{,}536x^{38}y^{45}z^{80})$$

673. $\dfrac{(2x^23y^3z^4)(8x^56y^3z^2)}{(6x^2y^23z^3)}$

$$\frac{(2\times3\times8\times6x^{2+5}y^{3+3}z^{4+2})}{(6x^2y^23z^3)}$$

$$\frac{(288x^7y^6z^6)}{(6x^2y^23z^3)}$$

$$288\div18x^{7-2}y^{6-2}z^{6-3}$$

$$16x^5y^4z^3$$

674. $\dfrac{(6x^35y^4z^6)(4x^43y^8z^2)}{(3x^32y^78z)}$

$$\frac{(6 \times 5 \times 4 \times 3x^{3+4}5y^{4+8}z^{6+2})}{(3x^3 2y^7 8z)}$$

$$\frac{(360x^7 y^{12} z^8)}{(3x^3 2y^7 8z)}$$

$$360 \div 48x^{7-3} y^{12-7} z^{8-1}$$

$$\frac{15}{2} x^4 y^5 z^7$$

675. $\frac{(x^3 6y^3 5z^2)(4x^4 y^3 3z^6)}{(x^8 8y^7 2z^3)}$

$$\frac{(6 \times 5 \times 4 \times 3x^{3+4} y^{3+3} z^{2+6})}{(x^8 8y^7 2z^3)}$$

$$\frac{(360x^7 y^6 z^8)}{(x^8 8y^7 2z^3)}$$

$$360 \div 16x^{7-8} y^{6-7} z^{8-3}$$

$$22\frac{1}{2} x^{-1} y^{-1} z^5$$

676. $\frac{(5x^8 6y^9 4z^6)(x^5 y^6 z^6)}{(2x^2 y^3 3z^2)}$

$$\frac{(5 \times 6 \times 4x^{8+5} y^{9+6} z^{6+6})}{(2x^2 y^3 3z^2)}$$

$$\frac{(120x^{13} y^{15} z^{12})}{(2x^2 y^3 3z^2)}$$

$$120 \div 6x^{13-2} y^{15-3} z^{12-2}$$

$$20x^{11} y^{12} z^{10}$$

677. $\frac{(5x^4 3y^2 2z^7)(6x^3 4y^8 z^6)}{(8x^2 2y^3 3z^2)}$

$$\frac{(5 \times 3 \times 2 \times 6 \times 4x^{4+3} y^{2+8} z^{7+6})}{(8x^2 2y^3 3z^2)}$$

$$\frac{(720x^7 y^{10} z^{13})}{(8x^2 2y^3 3z^2)}$$

$$720 \div 48x^{7-2} y^{10-3} z^{13-2}$$

$$15x^5 y^7 z^{11}$$

678. $\frac{(x^2 4y^6 2z^8)(5x^7 6y^4 z^2)^2}{(2x^6 2y^4 3z^3)}$

$$\frac{(x^24y^62z^8)(5x^76y^4z^2)^2}{(2x^62y^43z^3)}$$

$$\frac{(x^24y^62z^8)(5^2x^{7\times2}6^2y^{4\times2}z^{2\times2})}{(2x^62y^43z^3)}$$

$$\frac{(x^24y^62z^8)(25x^{14}36y^8z^4)}{(2x^62y^43z^3)}$$

$$\frac{(4\times2\times25\times36x^{2+14}y^{6+8}z^{8+4})}{(2x^62y^43z^3)}$$

$$\frac{(7{,}200x^{16}y^{14}z^{12})}{(2x^62y^43z^3)}$$

$$7{,}200\div12x^{16-6}y^{14-4}z^{12-3}$$

$$600x^{10}y^{10}z^9$$

679. $\dfrac{(x^3y^43z^2)(5x^66y^48z^8)}{(4x^46y^6z^9)}$

$$\frac{(3\times5\times6\times8x^{3+6}y^{4+4}z^{2+8})}{(4x^46y^6z^9)}$$

$$\frac{(720x^9y^8z^{10})}{(4x^46y^6z^9)}$$

$$720\div24x^{9-4}y^{8-6}z^{10-9}$$

$$30x^5y^2z$$

680. $\dfrac{(x^26y^25z^3)(3x^32y^4z^5)}{(2x^74y^82z^7)}$

$$\frac{(6\times5\times3\times2x^{2+3}y^{2+4}z^{3+5})}{(2x^74y^82z^7)}$$

$$\frac{(180x^5y^6z^8)}{(2x^74y^82z^7)}$$

$$180\div16x^{5-7}y^{6-8}z^{8-7}$$

$$11\frac{1}{4}x^{-2}y^{-2}z$$

681. $\dfrac{(6x^95y^4z^8)(x^3y^6z^2)}{(3x^24y^32z^5)}$

$$\frac{(6\times5x^{9+3}y^{4+6}z^{8+2})}{(3x^24y^32z^5)}$$

$$\frac{(30x^{12}y^{10}z^{10})}{(3x^24y^32z^5)}$$

$$\frac{(30x^{12}y^{10}z^{10})}{(3x^24y^32z^5)}$$

$$30 \div 24x^{12-2}y^{10-3}z^{10-5}$$

$$1\frac{1}{4}x^{10}y^7z^5$$

682. $\frac{(6x^63y^25z^4)(x^32y^7z^8)}{(2x^3y^4z^5)}$

$$\frac{(6 \times 3 \times 5 \times 2x^{6+3}y^{2+7}z^{4+8})}{(2x^3y^4z^5)}$$

$$\frac{(180x^9y^9z^{12})}{(2x^3y^4z^5)}$$

$$180 \;:\; 2x^{9-3}y^{9\;-\;4}z^{12-5}$$

$$90x^6y^5z^7$$

683. $\frac{(x^44y^66z^2)(5x^43y^32z^6)}{(6x^72y^2z^8)}$

$$\frac{(4 \times 6 \times 5 \times 3 \times 2x^{4+4}y^{6+3}z^{2+6})}{(6x^72y^2z^8)}$$

$$\frac{(720x^8y^9z^8)}{(6x^72y^2z^8)}$$

$$720 \div 12x^{8-7}y^{9-2}z^{8-8}$$

$$60xy^7$$

684. $\frac{(3x^2y^72z^5)(x^6y^94z^4)}{(3x^34y^24z^7)}$

$$\frac{(3 \times 2 \times 4x^{2+6}y^{7+9}z^{5+4})}{(3x^34y^24z^7)}$$

$$\frac{(24x^8y^{16}z^9)}{(3x^34y^24z^7)}$$

$$24 \div 48x^{8-3}y^{16-2}z^{9-7}$$

$$\frac{1}{2}x^5y^{14}z^2$$

685. $\frac{(8x^2y^72z^5)(4x^3y^95z^6)}{(2x^8y^24z^4)}$

$$\frac{(8 \times 2 \times 4 \times 5x^{2+3}y^{7+9}z^{5+6})}{(2x^8y^24z^4)}$$

$$\frac{(320x^5y^{16}z^{11})}{(2x^8y^24z^4)}$$

$$320 \div 8x^{5-8}y^{16-2}z^{11-4}$$

$$40x^{-3}y^{14}z^7$$

686. $\dfrac{(x^65y^54z^4)(2x^23y^4z^5)}{(2x^2y6z^2)}$

$$\frac{(5 \times 4 \times 2 \times 3x^{6+2}y^{5+4}z^{4+5})}{(2x^2y6)}$$

$$\frac{(120x^8y^9z^9)}{(2x^2y6z^2)}$$

$$120 \div 12x^{8-2}y^{9-1}z^{9-2}$$

$$10x^6y^8z^7$$

687. $\dfrac{(6x^25y^3z^6)(4x^2y^4z^8)}{(x^72y^53z^9)}$

$$\frac{(6 \times 5 \times 4x^{2+2}y^{3+4}z^{6+8})}{(x^72y^53z^9)}$$

$$\frac{(120x^4y^7z^{14})}{(x^72y^53z^9)}$$

$$120 \div 6x^{4-7}y^{7-5}z^{14-9}$$

$$20x^{-3}y^2z^5$$

688. $\dfrac{(3x^22y^6z^4)(x^55y^7z)}{(8x^87y^3z^9)}$

$$\frac{(3 \times 2 \times 5x^{2+5}y^{6+7}z^{4+1})}{(8x^87y^3z^9)}$$

$$\frac{(30x^7y^{13}z^5)}{(8x^87y^3z^9)}$$

$$30 \div 56x^{7-8}y^{13-3}z^{5-9}$$

$$\frac{15}{28}x^{-1}y^{10}z^{-4}$$

689. $\dfrac{(7x5y^4z^2)(3x^5y^87z^6)}{(7x^7y^6z^4)^2}$

$$\frac{(7x5y^4z^2)(3x^5y^87z^6)}{7^2x^{7\times2}y^{6\times2}z^{4\times2}}$$

$$\frac{(7x5y^4z^2)(3x^5y^87z^6)}{49x^{14}y^{12}z^8}$$

$$\frac{(7\times5\times3\times7x^{1+5}y^{4+8}z^{2+6}))}{49x^{14}y^{12}z^8}$$

$$\frac{735x^6y^{12}z^8}{49x^{14}y^{12}z^8}$$

$$735\div49x^{6-14}y^{12-1}\ z^{8-8}$$

$$15x^{-8}$$

690. $\frac{(x^95y^82z^2)(5x^33y^5z^4)}{(x^6y^4z^7)}$

$$\frac{(5\times2\times5\times3x^{9+3}y^{8+5}z^{2+4})}{(x^6y^4z^7)}$$

$$\frac{(150x^{12}y^{13}z^6)}{(x^6y^4z^7)}$$

$$150x^{12-6}y^{13-4}z^{6-7}$$

$$150x^6y^9z^{-1}$$

691. $\frac{(x^54y^45z^2)(3x^3y^8z^6)}{(2xy^7z^3)}$

$$\frac{(4\times5\times3x^{5+3}y^{4+8}z^{2+6})}{(2xy^7z^3)}$$

$$\frac{(60x^8y^{12}z^8)}{(2xy^7z^3)}$$

$$60\div2x^{8-1}y^{12-7}z^{8-3}$$

$$30x^7y^5z^5$$

692. $\frac{(x^28y^86z^3)(5x^54y^4z^6)}{(2x^94y^62z^7)}$

$$\frac{(8\times6\times5\times4x^{2+5}y^{8+4}z^{3+6})}{(2x^94y^64z^7)}$$

$$\frac{(960x^7y^{12}z^9)}{(2\times4\times2x^9y^6z^7)}$$

$$\frac{(960x^7y^{12}z^9)}{(16x^9y^6z^7)}$$

$$960 \div 16x^{7-9}y^{12-6}z^{9-7}$$

$$60x^{-2}y^6z^2$$

693. $\dfrac{(x^73y^4z^3)(7x^84y^6z^5)}{(5x^82y^36z^2)}$

$$\frac{(3 \times 7 \times 4x^{7+8}y^{4+6}z^{3+5})}{(5x^82y^36z^2)}$$

$$\frac{(84x^{15}y^{10}z^8)}{(5x^82y^36z^2)}$$

$$\frac{(84x^{15}y^{10}z^8)}{(5 \times 2 \times 6x^8y^3z^2)}$$

$$84 \div 60x^{15-8}y^{10-3}z^{8-2}$$

$$1\frac{2}{5}x^7y^7z^6$$

694. $\dfrac{(8x^9y^75z^4)^2(x^82y^34z^6)}{(x^28y^9z^4)}$

$$\frac{(8^2x^{9\times2}y^{7\times2}5^2z^{4\times2})(x^82y^34z^6)}{(x^28y^9z^4)}$$

$$\frac{(64x^{18}y^{14}25z^8)(x^82y^34z^6)}{(x^28y^9z^4)}$$

$$\frac{(64 \times 25 \times 2 \times 4x^{18+8}y^{14+3}z^{8+6})}{(x^28y^9z^4)}$$

$$\frac{(12{,}800x^{26}y^{17}z^{14})}{(x^28y^9z^4)}$$

$$12{,}800 \div 8x^{26-2}y^{17-9}z^{14-4}$$

$$1{,}600x^{24}y^8z^{10}$$

695. $\dfrac{(x^75y^52z^6)(3x^34y^9z^9)}{(x^4y8z^8)}$

$$\frac{(5 \times 2 \times 3 \times 4x^{7+3}y^{5+9}z^{6+9})}{(x^4y8z^8)}$$

$$\frac{(120x^{10}y^{14}z^{15})}{(x^4y8z^8)}$$

$$120 \div 8x^{10-4}y^{14-1}z^{15-8}$$

$$15x^6y^{13}z^7$$

696. $\dfrac{(8x^8y^2 5z^5)(3x^6 2y^7 4z^9)}{(x^3y^3z^4)}$

$$\dfrac{(8 \times 5 \times 3 \times 2 \times 4x^{8+6}y^{2+7}z^{5+9})}{(x^3y^3z^4)}$$

$$\dfrac{(960x^{14}y^9z^{14})}{(x^3y^3z^4)}$$

$$960x^{14-3}y^{9-3}z^{14-4}$$

$$960x^{11}y^6z^{10}$$

697. $\dfrac{(x^6 3y^4z^5)(2x^6 4y^9z^4)}{(2x^7y^4 6z^8)}$

$$\dfrac{(3 \times 2 \times 4x^{6+6}y^{4+9}z^{5+4})}{(2x^7y^4 6z^8)}$$

$$\dfrac{(24x^{12}y^{13}z^9)}{(2x^7y^4 6z^8)}$$

$$24 \div 12x^{12-7}y^{13-4}z^{9-8}$$

$$2x^5y^9z$$

698. $\dfrac{(x^8y^4 3z)(4x^3y^5z^8)}{(2x^7 8y^6z^9)}$

$$\dfrac{(3 \times 4x^{8+3}y^{4+5}z^{1+8})}{(2x^7 8y^6z^9)}$$

$$\dfrac{(12x^{11}y^9z^9)}{(2x^7 8y^6z^9)}$$

$$12 \div 16x^{11-7}y^{9-6}z^{9-9}$$

$$\dfrac{3}{4}x^4y^3$$

699. $\dfrac{(4x^7 2y^3 6z^5)(x^2 2y^3z^6)}{(8x^6 3y^4z^8)}$

$$\dfrac{(4 \times 2 \times 6 \times 2x^{7+2}y^{3+3}z^{5+6})}{(8x^6 3y^4z^8)}$$

$$\dfrac{(96x^9y^6z^{11})}{(8x^6 3y^4z^8)}$$

$$96 \div 24x^{9-6}y^{6-4}z^{11-8}$$

$$4x^3y^2z^3$$

700. $\dfrac{(x^6y^5z^2)(72x^9y^3z^8)}{(2x^4 3y^7 2z^2)^2}$

$$\frac{(72x^{6+9}y^{5+3}z^{2+8})}{(2^2x^{4\times2}3^2y^{7\times2}2^2z^{2\times2})}$$

$$\frac{(72x^{15}y^8z^{10})}{(144x^8y^{14}z^4)}$$

$$72 \div 144x^{15-8}y^{8-14}z^{10-4}$$

$$\frac{1}{2}x^7y^{-6}z^6$$

701. $\dfrac{(2x^6y^2z^5)(4x^7y^32z)}{(6x^43y^9z^8)}$

$$\frac{(2x^6y^2z^5)(4x^7y^32z)}{(6x^43y^9z^8)}$$

$$\frac{(2 \times 4 \times 2x^{6+7}y^{2+3}z^{5+1})}{(6x^43y^9z^8)}$$

$$\frac{(16x^{13}y^5z^6)}{(6x^43y^9z^8)}$$

$$16 \div 18x^{13-4}y^{5-9}z^{6-8}$$

$$\frac{8}{9}x^9y^{-4}z^{-2}$$

702. $\dfrac{(x^98y^6z^3)(3x^52y^54z^9)}{(8x^8y^4z^3)}$

$$\frac{(8 \times 3 \times 2 \times 4x^{9+5}y^{6+5}z^{3+9})}{(8x^8y^4z^3)}$$

$$\frac{(192x^{14}y^{11}z^{12})}{(8x^8y^4z^3)}$$

$$192 \div 8x^{14-8}y^{11-4}z^{12-3}$$

$$24x^6y^7z^9$$

703. $\dfrac{(x^8y^56z^2)(4x^3y^93z^7)}{(x^4y^66z^2)}$

$$\frac{(6 \times 4 \times 3x^{8+3}y^{5+9}z^{2+7})}{(x^4y^66z^2)}$$

$$\frac{(72x^{11}y^{14}z^9)}{(x^4y^66z^2)}$$

$$72 \div 6x^{11-4}y^{14-6}z^{9-2}$$

$$12x^7y^8z^7$$

704. $\dfrac{(x^9 2y^3 z^6)(3x^5 y^4 z^7)}{(x^2 y6z^4)}$

$$\frac{(x^9 2y^3 z^6)(3x^5 y^4 z^7)}{(x^2 y6z^4)}$$

$$\frac{(2 \times 3 \times 4x^{9+5} y^{3+4} z^{6+7})}{(x^2 y6z^4)}$$

$$\frac{(24x^{14} y^7 z^{13})}{(x^2 y6z^4)}$$

$$24 \div 6x^{14-2} y^{7-1} z^{13-4}$$

$$4x^{12} y^6 z^9$$

705. $\dfrac{(x^4 y^3 z^6)(2x^2 y^5 4z^5)}{(x^3 6y^7 z^8)}$

$$\frac{(2 \times 4x^{4+2} y^{3+5} z^{6+5})}{(x^3 6y^7 z^8)}$$

$$\frac{(8x^6 y^8 z^{11})}{(x^3 6y^7 z^8)}$$

$$8 \div 6x^{6-3} y^{8-7} z^{11-8}$$

$$1\frac{1}{3} x^3 y z^3$$

706. $\dfrac{(7x8y^4 z^5)(4x^6 3y^9 z^2)}{(6x^7 y^6 2z^8)}$

$$\frac{(7 \div 8 \div 4 \div 3x^{1+6} y^{4+9} z^{5+2})}{(6x^7 y^6 2z^8)}$$

$$\frac{(672x^7 y^{13} z^7)}{(6x^7 y^6 2z^8)}$$

$$672 \div 12x^{7-7} y^{13-6} z^{7-8}$$

$$56y^7 z^{-1}$$

707. $\dfrac{(8x^2 3y^9 2z^6)(7x^5 2y^3 z^3)}{(x^8 y^4 6z^7)}$

$$\frac{(8 \times 3 \times 2 \times 7 \times 2x^{2+5} y^{9+3} z^{6+3})}{(x^8 y^4 6z^7)}$$

$$\frac{(672x^7 y^{12} z^9)}{(x^8 y^4 6z^7)}$$

$$672 \div 6x^{7-8} y^{12-4} z^{9-7}$$

$$112x^{-1}y^8z^2$$

708. $\dfrac{(x^34y^5z^2)(3x2y^78z^5)}{(x^66y^9z^4)}$

$$\frac{(4 \times 3 \times 2 \times 8x^{3+1}y^{5+7}z^{2+5})}{(x^66y^9z^4)}$$

$$\frac{(192x^4y^{12}z^7)}{(x^66y^9z^4)}$$

$$32x^{-2}y^3z^3$$

709. $\dfrac{(x^5y^2z^7)(4x^37y^5z^8)}{(3x^42y^4z^9)^3}$

$$\frac{(4 \times 7x^{5+3}y^{2+5}z^{7+8})}{3^3x^{4\times3}2^3y^{4\times3}z^{9\times3}}$$

$$\frac{(28x^8y^7z^{15})}{27x^{12}8y^{12}z^{27}}$$

$$28 \div 216x^{8-12}y^{7-12}z^{15-27}$$

$$\frac{7}{54}x^{-4}y^{-5}z^{-1}$$

710. $\dfrac{(8x^93y^57z^3)(4x^7yz^3)}{(2x^46y^6z^6)}$

$$\frac{(8 \times 3 \times 7 \times 4x^{9+7}y^{5+1}z^{3+3})}{(2x^46y^6z^6)}$$

$$\frac{(672x^{16}y^6z^6)}{(2x^46y^6z^6)}$$

$$672 \div 12x^{16-4}y^{6-6}z^{6-6}$$

$$56x^{12}$$

711. $\dfrac{(x^83y^84z^3)(x^3y^5z^2)}{(6x^2y^42z^6)}$

$$\frac{(3 \times 4x^{8+3}y^{8+5}z^{3+2})}{(6x^2y^42z^6)}$$

$$\frac{(12x^{11}y^{13}z^5)}{(6x^2y^42z^6)}$$

$$12 \div 12x^{11-2}y^{13-4}z^{5-6}$$

$$x^9y^9z^{-1}$$

712. $\dfrac{(x^76y^4z^5)(3x^64y^47z^9)}{(8x^5y^9z)}$

$$\frac{(6 \times 3 \times 4 \times 7x^{7+6}y^{4+4}z^{5+9})}{(8x^5y^9z)}$$

$$\frac{(504x^{13}y^8z^{14})}{(8x^5y^9z)}$$

$$504 \div 8x^{13-5}y^{8-9}z^{14-1}$$

$$63x^8y^{-1}z^{13}$$

713. $\frac{(3x^8y^32z^6)(7x^54y^2z^7)}{(x^46y^8z^3)}$

$$\frac{(3 \times 2 \times 7 \times 4x^{8+5}y^{3+2}z^{6+7})}{(x^46y^8z^3)}$$

$$\frac{(168x^{13}y^5z^{13})}{(x^46y^8z^3)}$$

$$168 \div 6x^{13-4}y^{5-8}z^{13-3}$$

$$28x^9y^{-3}z^{10}$$

714. $\frac{(6x^23y^74z^4)(2x^6y^5z^3)}{(x^88y^2z^3)}$

$$\frac{(6 \times 3 \times 4 \times 2x^{2+6}y^{7+5}z^{4+3})}{(x^88y^2z^3)}$$

$$\frac{(144x^8y^{12}z^7)}{(x^88y^2z^3)}$$

$$144 \div 8x^{8-8}y^{12-2}z^{7-3}$$

$$18y^{10}z^4$$

715. $\frac{(7x^34y^92z^5)(x^3y^73z^5)}{(x^28y^4z^6)}$

$$\frac{(7 \times 4 \times 2 \times 3x^{3+3}y^{9+7}z^{5+5})}{(x^28y^4z^6)}$$

$$\frac{(168x^6y^{16}z^{10})}{(x^28y^4z^6)}$$

$$168 \div 8x^{6-2}y^{16-4}z^{10-6}$$

$$21x^4y^{12}z^4$$

716. $\frac{(6x^8y^4z^4)(2x4y^2z^4)}{(3x^6y^58z^7)}$

$$\frac{(6 \times 2 \times 4x^{8+1}y^{4+2}z^{4+4})}{(3x^6y^58z^7)}$$

$$\frac{(48x^9y^6z^8)}{(3x^6y^58z^7)}$$

$$48 \div 24x^{9-6}y^{6-5}z^{8-7}$$

$$2x^3yz$$

717. $\frac{(x^3y^22z^9)(7x^54y^3z^7)}{(3x^6y^46z^8)}$

$$\frac{(2 \times 7 \times 4x^{3+5}y^{2+3}z^{9+7})}{(3x^6y^46z^8)}$$

$$\frac{(56x^8y^5z^{16})}{(3x^6y^46z^8)}$$

$$56 \div 18x^{8-6}y^{5-4}z^{16-8}$$

$$\frac{28}{9}x^2yz^8$$

718. $\frac{(x^72y^24z^8)(x^37y^2z^6)}{(2x^47y^5z)}$

$$\frac{(2 \times 4 \times 7x^{7+3}y^{2+2}z^{8+6})}{(2x^47y^5z)}$$

$$\frac{(56x^{10}y^4z^{14})}{(2x^47y^5z)}$$

$$(56 \div 14x^{10-4}y^{4-5}z^{14-1})$$

$$4x^6y^{-1}z^{13}$$

719. $\frac{(x^23y^66z^4)(7x^92y^3z^3)}{(4x^52y^4z^7)^2}$

$$\frac{(3 \times 6 \times 7 \times 2x^{2+9}y^{6+3}z^{4+3})(x^9y^3z^3)}{(4^2x^{5\times2}2^2y^{4\times2}z^{7\times2})}$$

$$\frac{(252x^{11}y^9z^7)}{(16x^{10}4y^8z^{14})}$$

$$252 \div 64x^{11-10}y^{9-8}z^{7-14}$$

$$\frac{63}{16}xyz^{-7}$$

720. $\frac{(4x^32y^27z^4)(4x^73y^8z^6)}{(4x^5y^22z^3)}$

$$\frac{(4 \times 2 \times 7 \times 4 \times 3x^{3+7}y^{2+8}z^{4+6})}{(4x^5y^22z^3)}$$

$$\frac{(672x^{10}y^{10}z^{10})}{(4x^5y^22z^3)}$$

$$672 \div 8x^{10-5}y^{10-2}z^{10-3}$$

$$84x^5y^8z^7$$

721. $\frac{(5x^42y^23z^4)+(x^94y^34z^8)}{(2x^23y^6z^2)}$

$$\frac{(5 \times 2 \times 3x^4y^2z^4)}{(2 \times 3x^2y^6z^2)} + \frac{(4 \times 4x^9y^3z^8)}{(2 \times 3x^2y^6z^2)}$$

$$(30 \div 6x^{4-2}y^{2-6}z^{4-2}) + (16 \div 6x^{9-2}y^{3-6}z^{8-2})$$

$$5x^2y^{-4}z^2 + \frac{8}{3}x^7y^{-3}z^6$$

722. $\frac{(5x^4y^63z^2)+(4x^410y^4z^2)}{(5x^28y^3z^8)}$

$$\frac{(5 \times 3x^4y^6z^2)}{(5 \times 8x^2y^3z^8)} + \frac{(4 \times 10x^4y^4z^2)}{(5 \times 8x^2y^3z^8)}$$

$$15 \div 40x^{4-2}y^{6-3}z^{2-8} + 40 \div 40x^{4-2}y^{4-3}z^{2-8}$$

$$\frac{3}{8}x^2y^3z^{-6} + x^2yz^{-6}$$

723. $\frac{(2x^44y^4z^3)+(6x^25y^5z^7)}{(5x^52y^2z^5)}$

$$\frac{(2 \times 4x^4y^4z^3)}{(5 \times 2x^5y^2z^5)} + \frac{(6 \times 5x^2y^5z^7)}{(5 \times 2x^5y^2z^5)}$$

$$8 \div 10x^{4-5}y^{4-2}z^{3-5} + 30 \div 10x^{2-5}y^{5-2}z^{7-5}$$

$$\frac{4}{5}x^{-1}y^2z^{-2} + 3x^{-3}y^3z^2$$

724. $\frac{(5x^9y^47z^4)+(4x^2y^65z)}{(5x^2y^44z^3)}$

$$\frac{(5 \times 7x^9y^4z^4)}{(5 \times 4x^2y^4z^3)} + \frac{(4 \times 5x^2y^6z)}{(5 \times 4x^2y^4z^3)}$$

$$(35 \div 20x^{9-2}y^{4-4}z^{4-3}) + (20 \div 20y^{6-4}z^{1-3})$$

$$\frac{7}{4}x^7z + y^2z^{-2}$$

725. $\frac{(5x8y^4z^2)+(x^64y^62z^8)}{(10\ 7y^22z^4)}$

$$\frac{(5 \times 8xy^4z^2)}{(10 \times 2y^2z^4)} + \frac{(4 \times 2x^6y^6z^8)}{(10 \times 2x^7y^2z^4)}$$

$$(40 \div 20xy^{4-2}z^{2-4}) + (8 \div 20x^{6-7}y^{6-2}z^{8-4})$$

$$2x^6y^2z^{-2} + \frac{2}{5}x^{-1}y^4z^4$$

726. $\frac{(6x^85y^54z^4)+(3x^3y^2z^6)}{(2x^2y^46z^9)}$

$$\frac{(6 \times 5 \times 4x^8y^5z^4)}{(2 \times 6x^2y^4z^9)} + \frac{(3x^3y^2z^6)}{(2 \times 6x^2y^4z^9)}$$

$$(120 \div 12x^{8-2}y^{5-4}z^{4-9}) + (3 \div 12x^{3-2}y^{2-4}z^{6-9})$$

$$10x^6yz^{-5} + \frac{1}{4}xy^{-2}z^{-3}$$

727. $\frac{(x^74y^3z^4)+(3x^66y2z^3)}{(x^55y^26z^3)}$

$$\frac{(4x^7y^3z^4)}{(5 \times 6x^5y^2z^3)} + \frac{(3 \times 6 \times 2x^6yz^3)}{(5 \times 6x^5y^2z^3)}$$

$$4 \div 30x^{7-5}y^{3-2}z^{4-3} + 36 \div 30x^{6-5}y^{1-2}z^{3-3}$$

$$\frac{2}{15}x^2yz + \frac{6}{5}xy^{-1}$$

728. $\frac{(7x^82y^5z^4)+(3x^44y^2z^2)}{(6x^7y^32z^6)}$

$$\frac{(7 \times 2x^8y^5z^4)}{(6 \times 2x^7y^3z^6)} + \frac{(3 \times 4x^4y^2z^2)}{(6 \times 2x^7y^3z^6)}$$

$$\frac{7}{6}xy^2z^{-2} + x^{-3}y^{-1}z^{-4}$$

729. $\frac{(x^47y^56z^2)+(2x3y^84z^2)}{(2x^2y^36z^6)^2}$

$$\frac{(7x^4y^56z^2) + (2x3y^84z^2)}{(2^2x^{2\times2}y^{3\times2}6^2z^{6\times2})}$$

$$\frac{(7x^4y^56z^2) + (2x3y^84z^2)}{(4 \times 36x^4y^6z^{12})}$$

$$\frac{(7x^4y^56z^2)}{(144x^4y^6z^{12})} + \frac{(2x3y^84z^2)}{(144x^4y^6z^{12})}$$

$$\frac{(7 \times 6x^4y^5z^2)}{(144x^4y^6z^{12})} + \frac{(2 \times 3 \times 4xy^8z^2)}{(144x^4y^6z^{12})}$$

$$\frac{21}{72}y^{-1}z^{-1} + \frac{1}{6}x^{-3}y^2z^{-10}$$

730. $\dfrac{(2x^8y^27z^2)+(4x^36y^6z^7)}{(2x^93y^3z^5)}$

$$\frac{(2 \times 7x^8y^2z^2)}{(2 \times 3x^9y^3z^5)} + \frac{(4 \times 6x^3y^6z^7)}{(2 \times 3x^9y^3z^5)}$$

$$21 \div 6x^{8-9}y^{2-3}z^{2-5} + 24 \div 6x^{3-9}y^{6-3}z^{7-5}$$

$$\frac{7}{3}x^{-1}y^{-1}z^{-3} + 4x^{-6}y^3z^2$$

731. $\dfrac{(4x^3y^65z^2)+(2x^2y2z^9)}{(3x^5y^76z^2)}$

$$\frac{(4 \times 5x^3y^6z^2)}{(3 \times 6x^5y^7z^2)} + \frac{(2 \times 2x^2yz^9)}{(3 \times 6x^5y^7z^2)}$$

$$20 \div 18x^{3-5}y^{6-7}z^{2-2} + 4 \div 18x^{2-5}y^{1-7}z^{9-2}$$

$$\frac{10}{9}x^{-2}y^{-1} + \frac{2}{9}x^{-3}y^{-6}z^7$$

Multiply the two terms below:

732. $(x+1)(x+2) = x^2 + 3x + 2$
733. $(x+2)(x+7) = x^2 + 9x + 14$
734. $(x-5)(x+3) = x^2 - 2x - 15$
735. $(x+8)(x-4) = x^2 + 4x - 32$
736. $(x-6)(x-2) = x^2 - 8x + 12$
737. $(x+2)(x+3) = x^2 + 5x + 6$
738. $(x+7)(x+1) = x^2 + 8x + 7$
739. $(x-3)(x-5) = x^2 - 8x + 15$
740. $(x+8)(x+4) = x^2 + 12x + 32$
741. $(x-9)(x+6) = x^2 - 3x - 54$
742. $(x+8)(x+2) = x^2 + 10x + 16$
743. $(x+4)(x+8) = x^2 + 12x + 32$
744. $(x+2)(x-8) = x^2 - 6x - 16$
745. $(x+5)(x+7) = x^2 + 12x + 35$
746. $(x-2)(x+8) = x^2 + 6x - 16$
747. $(x+6)(x+5) = x^2 + 11x + 30$
748. $(x-3)(x+5) = x^2 + 2x - 15$
749. $(x-1)(x+3) = x^2 + 2x - 3$
750. $(x+1)(x+9) = x^2 + 10x + 9$

751. $(x - 7)(x - 7) = x^2 - 14x + 49$

752. $(x + 8)(x + 4) = x^2 + 12x + 32$

753. $(x + 6)(x + 3) = x^2 + 9x + 18$

754. $(x + 2)(x + 4) = x^2 + 6x + 8$

755. $(x - 5)(x + 6) = x^2 + x - 30$

756. $(x - 9)(x - 2) = x^2 - 11x + 18$

757. $(x + 5)(x + 12) = x^2 + 17x + 60$

758. $(x + 6)(x + 3) = x^2 + 9x + 18$

759. $(x - 3)(x - 4) = x^2 - 7x + 12$

760. $(x + 5)(x - 2) = x^2 + 3x - 10$

761. $(x + 4)(x - 11) = x^2 - 7x - 44$

762. $(x + 10)(x + 1) = x^2 + 11x + 10$

763. $(x - 5)(x + 9) = x^2 + 4x - 45$

764. $(x + 3)(x + 3) = x^2 + 6x + 9$

765. $(x - 8)(x + 9) = x^2 + x - 72$

766. $(x + 8)(x - 7) = x^2 + x - 56$

767. $(x + 2)(x - 4) = x^2 - 2x - 8$

768. $(x + 9)(x + 8) = x^2 + 17x + 72$

769. $(x - 2)(x + 3) = x^2 + x - 6$

770. $(x + 6)(x - 5) = x^2 + x - 30$

771. $(x + 3)(x + 7) = x^2 + 10x + 21$

772. $(x - 3)(x + 2) = x^2 - x - 6$

773. $(x + 5)(x + 9) = x^2 + 14x + 45$

774. $(x + 3)(x - 2) = x^2 + x - 6$

775. $(x + 4)(x + 4) = x^2 + 8x + 16$

776. $(x - 2)(4x + 3) = 4x^2 - 5x - 6$

777. $(2x + 2)(x - 5) = 2x^2 - 8x - 10$

778. $(x - 1)(2x + 3) = 2x^2 + x - 3$

779. $(x - 8)(3x + 2) = 3x^2 - 22x - 16$

780. $(5x + 2)(x - 4) = 5x^2 - 18x - 8$

781. $(x + 5)(4x + 4) = 4x^2 + 24x + 20$

782. $(x + 3)(x + 9) = x^2 + 12x + 27$

783. $(2x - 7)(2x + 6) = 4x^2 - 2x - 42$

Solve for x:

784. $\sqrt{8x + 9} = 5$

$$\sqrt{8x + 9}^2 = 5^2$$

$$8x + 9 = 25$$
$$8x = 25 - 9 = 16$$
$$\frac{8x}{8} = \frac{16}{8}$$
$$x = 2$$

785. $\sqrt{12x + 9} = 9$

$$\sqrt{12x + 9}^2 = 9^2$$
$$12x + 9 = 81$$
$$12x = 81 - 9 = 72$$
$$\frac{12x}{12} = \frac{72}{12}$$
$$x = 6$$

786. $\sqrt{x + 9} = 7$

$$\sqrt{x + 9}^2 = 7^2$$
$$x + 9 = 49$$
$$x = 49 - 9 = 40$$
$$x = 40$$

787. $\sqrt{15x + 9} = 12$

$$\sqrt{15x + 9}^2 = 12^2$$
$$15x + 9 = 144$$
$$15x = 144 - 9 = 135$$
$$\frac{15x}{15} = \frac{135}{15}$$
$$x = 9$$

788. $\sqrt{5x - 14} = 6$

$$\sqrt{5x - 14}^2 = 6^2$$
$$5x - 14 = 36$$
$$5x = 36 + 14 = 50$$
$$\frac{5x}{5} = \frac{50}{5}$$

$$x = 10$$

789. $\sqrt{5x + 6} = 9$

$$\sqrt{5x + 6}^2 = 9^2$$
$$5x + 6 = 81$$
$$5x = 81 - 6 = 75$$
$$\frac{5x}{5} = \frac{75}{5}$$
$$x = 15$$

790. $\sqrt{9x + 10} = 1$

$$\sqrt{9x + 10}^2 = 1^2$$
$$9x + 10 = 1$$
$$9x = 1 - 10 = -9$$
$$\frac{9x}{9} = \frac{-9}{9}$$
$$x = -1$$

791. $\sqrt{4x - 19} = 11$

$$\sqrt{4x - 19}^2 = 11^2$$
$$4x - 19 = 121$$
$$4x = 121 + 19 = 140$$
$$\frac{4x}{4} = \frac{140}{4}$$
$$x = 35$$

792. $\sqrt{2x + 16} = 12$

$$\sqrt{2x + 16}^2 = 12^2$$
$$2x + 16 = 144$$
$$2x = 144 - 16 = 128$$
$$\frac{2x}{2} = \frac{128}{2}$$
$$x = 64$$

793. $\sqrt{4x + 13} = 5$

$$\sqrt{4x + 13}^2 = 5^2$$

$$4x + 13 = 25$$

$$4x = 25 - 13 = 12$$

$$\frac{4x}{4} = \frac{12}{4}$$

$$x = 3$$

794. $\sqrt{6x} = 12$

$$\sqrt{6x}^2 = 12^2$$

$$6x = 144$$

$$\frac{6x}{6} = \frac{144}{6}$$

$$x = 24$$

795. $\sqrt{4x + 13} = 1$

$$\sqrt{4x + 13}^2 = 1^2$$

$$4x + 13 = 1$$

$$4x = 1 - 13 = -12$$

$$\frac{4x}{4} = \frac{-12}{4}$$

$$x = -3$$

796. $\sqrt{8x + 9} = 13$

$$\sqrt{8x + 9}^2 = 13^2$$

$$8x + 9 = 169$$

$$8x = 169 - 9 = 160$$

$$\frac{8x}{8} = \frac{-160}{8}$$

$$x = 20$$

797. $\sqrt{16x + 20} = 10$

$$\sqrt{16x + 20}^2 = 10^2$$

$$16x + 20 = 100$$

$$16x = 100 - 20 = 80$$

$$\frac{16x}{16} = \frac{80}{16}$$

$$x = 5$$

798. $\sqrt{10x + 9} = 7$

$$\sqrt{10x + 9}^{\,2} = 7^2$$

$$10x + 9 = 49$$

$$10x = 49 - 9 = 40$$

$$\frac{10x}{10} = \frac{40}{10}$$

$$x = 4$$

799. $\sqrt{x + 4} = 2$

$$\sqrt{x + 4}^{\,2} = 2^2$$

$$x + 4 = 4$$

$$x = 4 - 4 = 0$$

$$x = 0$$

800. $\sqrt{3x + 1} = 8$

$$\sqrt{3x + 1}^{\,2} = 8^2$$

$$3x + 1 = 64$$

$$3x = 64 - 1 = 63$$

$$3x = 63$$

$$\frac{3x}{3} = \frac{63}{3}$$

$$x = 21$$

Answers to Chapter 8 Quadratic Equations

801. $x^2 + 2x + 1 = (x + 1)(x + 1)$

802. $x^2 + 3x + 2 = (x + 1)(x + 2)$

803. $x^2 + 4x + 3 = (x + 3)(x + 1)$

804. $x^2 + 4x + 4 = (x + 2)(x + 2)$

805. $x^2 + 5x + 6 = (x + 2)(x + 3)$

806. $x^2 + 7x + 12 = (x + 3)(x + 4)$

807. $x^2 + 6x + 5 = (x + 1)(x + 5)$

808. $x^2 + 10x + 25 = (x + 5)(x + 5)$

809. $x^2 + 7x + 12 = (x + 3)(x + 4)$

810. $x^2 + 9x + 14 = (x + 2)(x + 7)$

811. $x^2 + 8x + 16 = (x + 4)(x + 4)$

812. $x^2 + 7x + 12 = (x + 4)(x + 3)$

813. $x^2 - 1 = (x - 1)(x + 1)$

814. $x^2 + x - 2 = (x + 2)(x - 1)$

815. $x^2 - 4 = (x - 2)(x + 2)$

816. $x^2 - 2x - 3 = (x - 3)(x + 1)$

817. $x^2 + 2x - 8 = (x + 4)(x - 2)$

818. $x^2 + x - 2 = (x + 2)(x - 1)$

819. $x^2 + x - 6 = (x + 3)(x - 2)$

820. $x^2 - 6x + 9 = (x - 3)(x - 3)$

821. $x^2 + 6x + 5 = (x + 1)(x + 5)$

822. $x^2 - 2x + 1 = (x - 1)(x - 1)$

823. $x^2 + 9x + 18 = (x + 3)(x + 6)$

824. $x^2 + 11x + 28 = (x + 4)(x + 7)$

825. $x^2 + 3x - 10 = (x - 2)(x + 5)$

826. $x^2 + 8x + 7 = (x + 7)(x + 1)$

827. $x^2 - 3x - 54 = (x + 6)(x - 9)$

828. $x^2 - 4x - 45 = (x - 9)(x + 5)$

829. $x^2 + 10x + 9 = (x + 1)(x + 9)$

830. $x^2 - x - 56 = (x - 8)(x + 7)$

831. $x^2 + 12x + 27 = (x + 3)(x + 9)$

832. $x^2 + 6x - 40 = (x + 10)(x - 4)$

833. $x^2 + 12x + 32 = (x + 4)(x + 8)$

834. $x^2 + 14x + 45 = (x + 5)(x + 9)$

835. $x^2 + 2x - 8 = (x - 2)(x + 4)$

836. $x^2 - 9 = (x + 3)(x - 3)$

837. $x^2 + 8x + 7 = (x + 1)(x + 7)$

838. $x^2 + 10x + 16 = (x + 8)(x + 2)$

839. $x^2 + 2x - 35 = (x - 5)(x + 7)$

840. $x^2 + 5x - 24 = (x + 8)(x - 3)$

841. $x^2 + 11x + 18 = (x + 2)(x + 9)$

842. $x^2 + 12x + 20 = (x + 10)(x + 2)$

843. $x^2 + 4x - 32 = (x - 4)(x + 8)$

844. $x^2 + 12x + 27 = (x + 9)(x + 3)$

845. $x^2 + 10x + 21 = (x + 3)(x + 7)$

846. $x^2 + 16x + 64 = (x + 8)(x + 8)$

847. $x^2 - 7x - 30 = (x + 3)(x - 10)$

848. $x^2 + 11x + 28 = (x + 4)(x + 7)$

849. $x^2 + 9x + 14 = (x + 2)(x + 7)$

850. $x^2 + 15x + 50 = (x + 5)(x + 10)$

851. $x^2 - 18x + 80 = (x - 8)(x - 10)$

852. $x^2 + 10x + 9 = (x + 9)(x + 1)$

853. $x^2 - 14x + 49 = (x - 7)(x - 7)$

854. $x^2 + 18x + 81 = (x + 9)(x + 9)$

855. $x^2 + 17x + 60 = (x + 12)(x + 5)$

856. $x^2 - 5x + 6 = (x - 2)(x - 3)$

857. $x^2 + 9x + 8 = (x + 8)(x + 1)$

858. $x^2 + x - 56 = (x - 7)(x + 8)$

859. $x^2 + 10x + 9 = (x + 1)(x + 9)$

860. $x^2 + 3x - 10 = (x + 5)(x - 2)$

861. $x^2 + 8x + 12 = (x + 6)(x + 2)$

862. $x^2 + 5x + 6 = (x + 2)(x + 3)$

863. $x^2 + 12x + 35 = (x + 5)(x + 7)$

864. $x^2 + 7x - 18 = (x - 2)(x + 9)$

865. $x^2 - x - 12 = (x - 4)(x + 3)$

866. $x^2 - 3x - 28 = (x - 7)(x + 4)$

867. $x^2 - 36 = (x - 6)(x + 6)$

868. $x^2 + 10x + 16 = (x + 8)(x + 2)$

869. $x^2 - 3x - 10 = (x + 2)(x - 5)$

870. $x^2 - 2x - 15 = (x + 3)(x - 5)$

871. $x^2 - 16 = (x + 4)(x - 4)$

872. $2x^2 + x - 1 = (2x - 1)(x + 1)$

873. $2x^2 - 2 = 2(x + 1)(x - 1)$

874. $3x^2 + x - 2 = (3x - 2)(x + 1)$

875. $2x^2 + 6x + 4 = 2(x + 1)(x + 2)$
876. $x^2 - 9 = (x + 3)(x - 3)$
877. $4x^2 - 13x - 12 = (x - 4)(4x + 3)$
878. $9x^2 - 9 = 9(x - 1)(x + 1)$
879. $2x^2 - 8x - 10 = 2(x + 1)(x - 5)$
880. $3x^2 - 3 = 3(x - 1)(x + 1)$
881. $4x^2 + 14x + 6 = 2(2x + 1)(x + 3)$
882. $2x^2 + 4x - 16 = 2(x + 4)(x - 2)$
883. $x^2 + 9x + 18 = (x + 6)(x + 3)$
884. $3x^2 + 28x + 49 = (x + 7)(3x + 7)$
885. $3x^2 + 4x + 1 = (3x + 1)(x + 1)$
886. $x^2 - 3x - 28 = (x + 4)(x - 7)$
887. $2x^2 - x - 15 = (x - 3)(2x + 5)$
888. $5x^2 - 4x - 1 = (5x + 1)(x - 1)$
889. $9x^2 + 42x + 49 = (3x + 7)(3x + 7)$
890. $4x^2 - 4 = 4(x + 1)(x - 1)$
891. $x^2 + 12x + 27 = (x + 3)(x + 9)$
892. $8x^2 - 44x + 36 = 4(x - 1)(2x - 9)$
893. $2x^2 + 20x + 18 = 2(x + 9)(x + 1)$
894. $x^2 - 6x - 40 = (x + 4)(x - 10)$
895. $10x^2 + 25x + 15 = 5(x + 1)(2x + 3)$
896. $3x^2 + 19x - 14 = (3x - 2)(x + 7)$
897. $x^2 + 13x + 36 = (x + 4)(x + 9)$
898. $x^2 + x - 42 = (x - 6)(x + 7)$
899. $3x^2 + 15x + 12 = 3(x + 4)(x + 1)$
900. $x^2 + 2x - 35 = (x + 7)(x - 5)$
901. $3x^2 + 5x + 2 = (3x + 2)(x + 1)$
902. $x^2 + 6x - 16 = (x - 2)(x + 8)$
903. $9x^2 + 12x + 4 = (3x + 2)(3x + 2)$
904. $4x^2 + 12x + 9 = (2x + 3)(2x + 3)$
905. $10x^2 - 29x + 10 = (2x - 5)(5x - 2)$
906. $x^2 + 4x + 4 = (x + 2)(x + 2), x = -2$
907. $x^2 + 5x + 6 = (x + 2)(x + 3), x = -2, -3$
908. $x^2 + 11x + 30 = (x + 5)(x + 6), x = -5, -6$
909. $x^2 + 8x + 12 = (x + 2)(x + 6), x = -2, -6$
910. $x^2 + 11x + 24 = (x + 3)(x + 8), x = -3, -8$
911. $x^2 + 9x + 8 = (x + 8)(x + 1), x = 8, -1$
912. $x^2 - x - 12 = (x - 4)(x + 3), x = 4, -3$

913. $x^2 + x - 20 = (x + 5)(x - 4), x = -5, 4$

914. $x^2 - x - 90 = (x - 10)(x + 9), x = 10, -9$

915. $x^2 - x - 110 = (x + 10)(x - 11), x = -10, 11$

916. $x^2 - x - 30 = (x + 5)(x - 6), x = -5, 6$

917. $x^2 - 14x + 40 = (x - 4)(x - 10), x = 4, 10$

918. $x^2 + 22x + 120 = (x + 12)(x + 10), x = -12, -10$

919. $x^2 + 15x + 54 = (x + 6)(x + 9), x = -6, -9$

920. $x^2 - 5x - 24 = (x + 3)(x - 8), x = -3, 8$

921. $x^2 - 12x + 35 = (x - 5)(x - 7), x = 5, 7$

922. $x^2 - 2x - 8 = (x + 2)(x - 4), x = -2, 4$

923. $x^2 + 7x - 8 = (x + 8)(x - 1), x = -8, 1$

924. $x^2 + x - 110 = (x + 11)(x - 10), x = -11, 10$

925. $x^2 + 10x + 21 = (x + 3)(x + 7), x = -3, -7$

926. $x^2 + 3x - 40 = (x - 5)(x + 8), x = 5, -8$

927. $x^2 + 9x - 36 = (x - 3)(x + 12), x = 3, -12$

928. $x^2 - 7x - 60 = (x + 5)(x - 12), x = -5, 12$

929. $x^2 + 20x + 96 = (x + 12)(x + 8), x = -12, -8$

930. $x^2 + 8x - 33 = (x + 11)(x - 3), x = -11, 3$

931. $x^2 - 14x + 33 = (x - 11)(x - 3), x = 11, 3$

932. $x^2 + 5x - 84 = (x + 12)(x - 7), x = -12, 7$

933. $x^2 + 16x + 48 = (x + 4)(x + 12), x = -4, -12$

934. $x^2 + 10x - 75 = (x + 15)(x - 5), x = -15, 5$

935. $x^2 - 14x + 48 = (x - 6)(x - 8), x = 6, 8$

936. $x^2 + x - 6 = (x + 3)(x - 2), x = -3, 2$

937. $x^2 - 2x - 48 = (x + 6)(x - 8), x = -6, 8$

938. $x^2 - x - 12 = (x + 3)(x - 4), x = -3, 4$

939. $x^2 - x - 20 = (x + 4)(x - 5), x = -4, 5$

940. $x^2 - x - 12 = (x + 3)(x - 4), x = -3, 4$

941. $x^2 + x - 30 = (x - 5)(x + 6), x = 5, -6$

942. $x^2 - 6x - 72 = (x - 12)(x + 6), x = 12, -6$

943. $x^2 - 12x + 35 = (x - 7)(x - 5), x = 7, 5$

944. $x^2 + 11x + 10 = (x + 1)(x + 10), x = -1, -10$

945. $x^2 + 2x - 168 = (x - 12)(x + 14), x = 12, -14$

946. $x^2 - x - 132 = (x - 12)(x + 11), x = 12, -11$

947. $x^2 - 11x + 30 = (x - 6)(x - 5), x = 6, 5$

948. $x^2 + 3x - 10 = (x - 2)(x + 5), x = 2, -5$

949. $x^2 - 3x - 180 = (x + 12)(x - 15), x = -12, 15$

950. $x^2 + x - 6 = (x + 3)(x - 2), x = -3, 2$

951. $3x^2 + 4x - 4 = (3x - 2)(x + 2)$

$$3x - 2 = 0$$
$$3x = 2$$
$$\frac{3x}{3} = \frac{2}{3}$$
$$x = \frac{2}{3}, -2$$

952. $4x^2 + 21x + 5 = (x + 5)(4x + 1)$

$$4x + 1 = 0$$
$$4x = -1$$
$$\frac{4x}{4} = -\frac{1}{4}$$
$$x = -\frac{1}{4}, -5$$

953. $2x^2 + 8x - 10 = (x + 5)(2x - 2)$

$$2x - 2 = 0$$
$$2x = 2$$
$$\frac{2x}{2} = \frac{2}{2}$$
$$x = 1, -5$$

954. $4x^2 - 4 = (2x - 2)(2x + 2)$

$$2x - 2 = 0$$
$$2x = 2$$
$$\frac{2x}{2} = \frac{2}{2}$$
$$x = 1$$
$$2x + 2 = 0$$
$$2x = -2$$
$$\frac{2x}{2} = \frac{-2}{2}$$
$$x = 1, -1$$

955. $4x^2 + 13x + 3 = (4x + 1)(x + 3)$

$$4x + 1 = 0$$
$$4x = -1$$
$$\frac{4x}{4} = -\frac{1}{4}$$
$$x = -\frac{1}{4}, -3$$

956. $2x^2 + 19x + 35 = (x + 7)(2x + 5)$

$$2x + 5 = 0$$
$$2x = -5$$

$$\frac{2x}{2} = -\frac{5}{2}$$

$$x = -\frac{5}{2}, -7$$

957. $x^2 + 9x + 20 = (x + 5)(x + 4), x = -5, -4$

958. $6x^2 + 34x + 48 = (3x + 8)(2x + 6)$

$$3x + 8 = 0$$

$$\frac{3x}{3} = \frac{-8}{3}$$

$$x = -\frac{8}{3}$$

$$\frac{2x}{2} = \frac{-6}{2}$$

$$x = -3, -\frac{8}{3}$$

959. $4x^2 + 8x + 4 = (2x + 2)(2x + 2)$

$$2x + 2 = 0$$

$$2x = -2$$

$$\frac{2x}{2} = \frac{-2}{2}$$

$$x = -1$$

960. $6x^2 - 19x + 15 = (2x - 3)(3x - 5)$

$$2x - 3 = 0$$

$$2x = 3$$

$$\frac{2x}{2} = \frac{3}{2}$$

$$x = \frac{3}{2}$$

$$3x - 5 = 0$$

$$3x = 5$$

$$\frac{3x}{3} = \frac{5}{3}$$

$$x = \frac{3}{2}, \frac{5}{3}$$

961. $3x^2 + 20x + 25 = (x + 5)(3x + 5)$

$$3x + 5 = 0$$

$$3x = -5$$

$$\frac{3x}{3} = \frac{-5}{3}$$

$$x = -5, -\frac{5}{3}$$

962. $x^2 - 3x - 10 = (x + 2)(x - 5), x = -2, 5$

963. $x^2 + 11x + 18 = (x + 9)(x + 2), x = -9, -2$

964. $x^2 - 16x + 48 = (x - 4)(x - 12), x = 4, 12$

965. $10x^2 + 18x + 8 = (2x + 2)(5x + 4)$

$$2x + 2 = 0$$
$$2x = -2$$
$$\frac{2x}{2} = \frac{-2}{2}$$
$$x = -1$$
$$5x + 4 = 0$$
$$5x = -4$$
$$\frac{5x}{5} = \frac{-4}{5}$$
$$x = -1, -\frac{4}{5}$$

966. $x^2 + x - 132 = (x + 12)(x - 11), x = -12, 11$

967. $4x^2 + 10x + 6 = (2x + 3)(2x + 2)$

$$2x + 3 = 0$$
$$2x = -3$$
$$\frac{2x}{2} = -\frac{3}{2}$$
$$x = -\frac{3}{2}$$
$$2x + 2 = 0$$
$$2x = -2$$
$$\frac{2x}{2} = \frac{-2}{2}$$
$$x = -\frac{3}{2}, -1$$

968. $x^2 - 25 = (x - 5)(x + 5), x = 5, -5$

969. $6x^2 + 23x + 7 = (3x + 1)(2x + 7)$

$$3x + 1 = 0$$
$$3x = -1$$
$$\frac{3x}{3} = -\frac{1}{3}$$
$$x = -\frac{1}{3}$$
$$2x + 7 = 0$$
$$2x = -7$$
$$\frac{2x}{2} = \frac{-7}{2}$$

$$x = -\frac{1}{3}, -\frac{7}{2}$$

970. $6x^2 - 14x + 8 = (3x - 4)(2x - 2)$

$$3x - 4 = 0$$
$$3x = 4$$
$$\frac{3x}{3} = \frac{4}{3}$$
$$x = \frac{4}{3}$$
$$2x - 2 = 0$$
$$2x = 2$$
$$\frac{2x}{2} = \frac{2}{2}$$
$$x = \frac{4}{3}, 1$$

971. $x^2 = 25$

$$\sqrt{x^2} = \sqrt{25}$$
$$x = \pm 5$$

972. $x^2 = 36$

$$\sqrt{x^2} = \sqrt{36}$$
$$x = \pm 6$$

973. $x^2 = 100$

$$\sqrt{x^2} = \sqrt{100}$$
$$x = \pm 10$$

974. $x^2 = 64$

$$\sqrt{x^2} = \sqrt{64}$$
$$x = \pm 8$$

975. $x^2 = 121$

$$\sqrt{x^2} = \sqrt{121}$$
$$x = \pm 11$$

976. $4x^2 = 64$

$$\frac{4x^2}{4} = \frac{64}{4}$$
$$\sqrt{x^2} = \sqrt{16}$$
$$x = \pm 4$$

977. $3x^2 - 5 = 7$

$$3x^2 - 5 = 7$$
$$3x^2 = 7 + 5 = 12$$
$$\frac{3x^2}{3} = \frac{12}{3}$$
$$x^2 = 4$$
$$\sqrt{x^2} = \sqrt{4}$$
$$x = \pm 2$$

978. $10x^2 = 1{,}000$

$$\frac{10x^2}{10} = \frac{1{,}000}{10}$$
$$x^2 = 100$$
$$\sqrt{x^2} = \sqrt{100}$$

$$x = \pm 10$$

979. $x^2 = 10,000$

$$x^2 = 10,000$$

$$\sqrt{x^2} = \sqrt{10,000}$$

$$x = \pm 100$$

980. $5x^2 - 20 = 480$

$$5x^2 - 20 = 480$$

$$5x^2 = 480 + 20 = 500$$

$$\frac{5x^2}{5} = \frac{500}{5}$$

$$x^2 = 100$$

$$\sqrt{x^2} = \sqrt{100}$$

$$x = \pm 10$$

981. $2x^2 + 25 = 123$

$$2x^2 + 25 = 123$$

$$2x^2 = 123 - 25 = 98$$

$$\frac{2x^2}{2} = \frac{98}{2}$$

$$x^2 = 49$$

$$\sqrt{x^2} = \sqrt{49}$$

$$x = \pm 7$$

982. $3x^2 - 13 = 350$

$$11x^2 - 13 = 350$$

$$11x^2 = 350 + 13$$

$$11x^2 = 363$$

$$\frac{11x^2}{11} = \frac{363}{11}$$

$$x^2 = 121$$

$$\sqrt{x^2} = \sqrt{121}$$

$$x = \pm 11$$

983. $x^2 - 21 = 123$

$$x^2 - 21 = 123$$

$$x^2 = 123 + 21 = 144$$

$$x^2 = 144$$

$$\sqrt{x^2} = \sqrt{144}$$

$$x = \pm 12$$

984. $3x^2 - 30 = -3$

$$3x^2 - 30 = -3$$

$$3x^2 = -3 + 30$$

$$3x^2 = 27$$

$$\frac{3x^2}{3} = \frac{27}{3}$$

$$x^2 = 9$$

$$\sqrt{x^2} = \sqrt{9}$$

$$x = \pm 3$$

985. $3x^2 + 45 = 192$

$$3x^2 + 45 = 192$$

$$3x^2 = 192 - 45 = 147$$

$$\frac{3x^2}{3} = \frac{147}{3}$$

$$x^2 = 49$$

$$\sqrt{x^2} = \sqrt{49}$$

$$x = \pm 7$$

986. $9x^2 - 27 = 54$

$$9x^2 - 27 = 54$$

$$9x^2 = 54 + 27 = 81$$

$$\frac{9x^2}{9} = \frac{81}{9}$$

$$x^2 = 9$$

$$\sqrt{x^2} = \sqrt{9}$$

$$x = \pm 3$$

987. $2x^2 - 60 = 140$

$$2x^2 = 140 + 60 = 200$$

$$\frac{200x^2}{2} = \frac{200}{2}$$

$$x^2 = 100$$

$$\sqrt{x^2} = \sqrt{100}$$

$$x = \pm 10$$

988. $3x^2 + 16 = 124$

$$3x^2 = 124 - 16 = 108$$

$$\frac{3x^2}{3} = \frac{108}{3}$$

$$x^2 = 36$$

$$\sqrt{x^2} = \sqrt{36}$$

$$x = \pm 6$$

991. $3x^2 + 15x + 18$

989. $8x^2 - 23 = 9$

$$8x^2 = 9 + 23 = 32$$

$$\frac{8x^2}{8} = \frac{32}{8}$$

$$x^2 = 4$$

$$\sqrt{x^2} = \sqrt{4}$$

$$x = \pm 2$$

990. $x^2 - 45 = 211$

$$x^2 - 45 = 211$$

$$x^2 = 211 + 45 = 256$$

$$\sqrt{x^2} = \sqrt{256}$$

$$x = \pm 16$$

$$x = \frac{-b \pm \sqrt{b^2 - 4ac}}{2a}$$

$$x = \frac{-15 \pm \sqrt{15^2 - 4(3)(18)}}{2(3)}$$

$$x = \frac{-15 \pm \sqrt{225 - 216}}{6}$$

$$x = \frac{-15 \pm \sqrt{9}}{6} = \frac{-15 \pm 3}{6}, = -\frac{18}{6}, -\frac{12}{6} = -3, -2$$

992. $x^2 + 4x + 4$

$$x = \frac{-b \pm \sqrt{b^2 - 4ac}}{2a}$$

$$x = \frac{-4 \pm \sqrt{4^2 - 4(1)(4)}}{2(1)}$$

$$x = \frac{-4 \pm \sqrt{16 - 16}}{2}$$

$$x = \frac{-4}{2} = -2$$

993. $2x^2 + 19x + 35$

$$x = \frac{-b \pm \sqrt{b^2 - 4ac}}{2a}$$

$$x = \frac{-19 \pm \sqrt{19^2 - 4(2)(35)}}{2(2)}$$

$$x = \frac{-19 \pm \sqrt{361 - 280}}{4}$$

$$x = \frac{-19 \pm \sqrt{81}}{4}$$

$$x = \frac{-19 \pm 9}{4} = -7, -\frac{5}{2}$$

994. $5x^2 + 36x + 36$

$$x = \frac{-b \pm \sqrt{b^2 - 4ac}}{2a}$$

$$x = \frac{-36 \pm \sqrt{36^2 - 4(5)(36)}}{2(5)}$$

$$x = \frac{-36 \pm \sqrt{1,296 - 720}}{10}$$

$$x = \frac{-36 \pm \sqrt{576}}{10}$$

$$x = \frac{-36 \pm 24}{10}$$

$$x = \frac{-36 \pm 24}{10} = -6, -\frac{6}{5}$$

995. $4x^2 + 25x + 25$

$$x = \frac{-b \pm \sqrt{b^2 - 4ac}}{2a}$$

$$x = \frac{-25 \pm \sqrt{25^2 - 4(4)(25)}}{2(4)}$$

$$x = \frac{-25 \pm \sqrt{625 - 400}}{8}$$

$$x = \frac{-25 \pm \sqrt{225}}{8}$$

$$x = \frac{-25 \pm 15}{8} = -\frac{5}{4}, -5$$

996. $6x^2 + 43x + 42$

$$x = \frac{-b \pm \sqrt{b^2 - 4ac}}{2a}$$

$$x = \frac{-43 \pm \sqrt{43^2 - 4(6)(42)}}{2(6)}$$

$$x = \frac{-43 \pm \sqrt{841}}{12}$$

$$x = \frac{-43 \pm 29}{12} = -6, -\frac{7}{6}$$

997. $2x^2 + 11x + 12$

$$x = \frac{-b \pm \sqrt{b^2 - 4ac}}{2a}$$

$$x = \frac{-11 \pm \sqrt{11^2 - 4(2)(12)}}{2(2)}$$

$$x = \frac{-11 \pm \sqrt{121 - 96}}{4}$$

$$x = \frac{-11 \pm \sqrt{25}}{4}$$

$$x = \frac{-11 \pm 5}{4} = -\frac{3}{2}, -4$$

998. $3x^2 + 14x + 8$

$$x = \frac{-b \pm \sqrt{b^2 - 4ac}}{2a}$$

$$x = \frac{-14 \pm \sqrt{14^2 - 4(3)(8)}}{2(3)}$$

$$x = \frac{-14 \pm \sqrt{196 - 96}}{6}$$

$$x = \frac{-14 \pm \sqrt{100}}{6}$$

$$x = \frac{-14 \pm 10}{6} = -4, -\frac{2}{3}$$

999. $x^2 + 8x + 15$

$$x = \frac{-b \pm \sqrt{b^2 - 4ac}}{2a}$$

$$x = \frac{-8 \pm \sqrt{8^2 - 4(1)(15)}}{2(1)}$$

$$x = \frac{-8 \pm \sqrt{64 - 60}}{2}$$

$$x = \frac{-8 \pm \sqrt{4}}{2}$$

$$x = \frac{-8 \pm 2}{2} = -5, -3$$

1000. $4x^2 + 17x + 18$

$$x = \frac{-b \pm \sqrt{b^2 - 4ac}}{2a}$$

$$x = \frac{-17 \pm \sqrt{17^2 - 4(4)(18)}}{2(4)}$$

$$x = \frac{-17 \pm \sqrt{289 - 288}}{8}$$

$$x = \frac{-17 \pm 1}{8} = -\frac{9}{4}, -2$$

Answer to Problem 1,001

1001. Why do we have to study Algebra, and how does Algebra help us in life?

Great question! One can consider that school exists for three reasons: 1-provide childcare (so parents can work), 2-prepare students for adult life, 3-prepare students for college.

It seems obvious that Algebra is not needed for childcare or for adult life. Therefore, one can only assume that Algebra is taught in order to prepare students for college. Presumably, to prepare students for a STEM (Science, Technology, Engineering and Math) course of study in college. Algebra is the basic building block for more advanced courses, such as Physics and Calculus.

We can therefore think of Algebra as a sort of "gateway" course; it's not needed for your future, but it is needed to keep your options open in the future.

While I can confirm that I have rarely, if ever, used Algebra in my adult life, I cannot say the same for Physics or Calculus. I have made rare usage of both disciplines on occasion, almost always when solving novel engineering problems – i.e., deriving new equations for new concepts that did not previously exist.

A lot has been discussed about how the current education system fails to prepare students for adult life. While students suffer through Algebra courses, little effort is made to teach them about the American legal system, healthcare, economics, buying a home or career development. It's no wonder that so many kids fail to achieve in high school if a clear line between their education and preparation for adult life is not established.

But Algebra does help in one aspect of adult life—teaching critical thinking and reasoning. Algebra problems are challenging. Solving 100 Algebra problems in one day is even more challenging. While the knowledge and experience gained may not be entirely helpful in life, the ability to focus on difficult problems and find solutions, over and over again, will most certainly help anyone later in life.

Therefore, you should think of the Algebra problems as meaningless little puzzles. But you are not learning Algebra when you solve the puzzles, you are really learning how to think critically and solve difficult problems, training and teaching your brain to focus and find solutions. Those who learn to excel at this will have the advantage of strong critical thinking capabilities to use for their entire life.

You're Done!

Congratulations! Easy, right? If you didn't skip any problems, then you have just completed more Algebra practice problems than most people will complete in their entire life.

In math, practice makes perfect. So, by now you should be close to perfect.

If you would like to reach me for any reason, you can reach me at josiahcoates@gmail.com.

Made in the USA
Las Vegas, NV
01 October 2021